Textbook of Chemistry

Textbook of Chemistry

Belle Caldwell

NY RESEARCH PRESS
New York

Published by NY Research Press
118-35 Queens Blvd., Suite 400,
Forest Hills, NY 11375, USA
www.nyresearchpress.com

Textbook of Chemistry
Belle Caldwell

International Standard Book Number: 978-1-63238-838-4 (Hardback)

Cataloging-in-Publication Data

Textbook of chemistry / Belle Caldwell.
p. cm.
Includes bibliographical references and index.
ISBN 978-1-63238-838-4
1. Chemistry. 2. Physical sciences. I. Caldwell, Belle.
QD31.3 .T49 2022
540--dc23

Contents

Preface

The scientific discipline which deals with the elements and compounds that are composed of atoms, molecules and ions is known as chemistry. Their composition, structure, behavior, properties and the changes they undergo during a reaction with other substances are studied under chemistry. It acts as a foundation in understanding the basic and applied scientific disciplines, such as ecology, botany, geology, pharmacology, etc. Some of the important aspects studied under this discipline are chemical bonding, chemical compounds, chemical reactions, acidity and basicity, chemical equilibrium, etc. Analytical chemistry, organic chemistry, inorganic chemistry, biochemistry, physical chemistry, nuclear chemistry, etc. are some of the sub-disciplines of this umbrella field. This textbook is a compilation of chapters that discuss the most vital concepts in the field of chemistry. Some of the diverse topics covered in this book address the varied branches that fall under this subject. This textbook will provide comprehensive knowledge to the readers.

Given below is the chapter wise description of the book:

Chapter 1- Chemistry is a branch of science that involves elements and compounds composed of atoms, molecules and ions. It focuses on their composition, behavior, properties, structures and the changes they make during a chemical reaction with other materials. This is an introductory chapter which will introduce briefly all the significant aspects of chemistry.

Chapter 2- The smallest unit of matter with the properties of chemical element is known as an atom. It comprises of a nucleus and one or more than one electron bound to the nucleus. Molecule is a group of atoms that is bounded together by the chemical bonds. This chapter has been carefully written to provide an easy understanding of the varied facets of atoms and molecules.

Chapter 3- The attraction between atoms, ions and molecules that facilitates the formation of chemical compounds is known as chemical bond. The various types of chemical bonds include covalent bond, ionic bond, metallic bond, etc. A chemical reaction is a process that transforms one set of chemical substances to another. The topics elaborated in this chapter will help in gaining a better perspective of chemical bonding and reactivity.

Chapter 4- Chemical kinetics is the domain of physical chemistry that studies the rates of chemical reactions. It focuses on rate laws and rate constants, reaction mechanism, collision theory, activation energy, etc. The chapter closely examines these key concepts of chemical kinetics to provide an extensive understanding of the subject.

Chapter 5- The study of the interrelation of heat and work with chemical reactions is known as chemical thermodynamics. In order to study chemical questions and the spontaneity of processes it involves laboratory measurements of thermodynamic properties and diverse mathematical methods. The diverse aspects of chemical thermodynamics in the current scenario have been thoroughly discussed in this chapter.

Chapter 6- Thermochemistry is a domain that studies heat energy related to chemical reactions and physical transformations. It involves the study of laws of chemistry, heat capacity, calorimetry, thermochemical equations, molar heat capacity, etc. All the diverse aspects of thermochemistry have been carefully analyzed in this chapter.

Chapter 7- Electrochemistry deals with the interaction between electrical energy and chemical change. Electrochemical cell is a tool via which electric current is produced from energy that is released by a spontaneous redox reaction. This chapter discusses in detail the theories related to electrochemistry.

Chapter 8- An acid is a molecule or an ion that can donate a proton as well as form a covalent bond with an electron pair. Bases are the substance that release hydroxide in an aqueous solution. ph is the scale that is used to identify acidic or basic nature of a water solution. This chapter has been carefully written to provide an easy understanding of the varied facets of acids and bases.

At the end, I would like to thank all those who dedicated their time and efforts for the successful completion of this book. I also wish to convey my gratitude towards my friends and family who supported me at every step.

Belle Caldwell

1

An Introduction to Chemistry

Chemistry is a branch of science that involves elements and compounds composed of atoms, molecules and ions. It focuses on their composition, behavior, properties, structures and the changes they make during a chemical reaction with other materials. This is an introductory chapter which will introduce briefly all the significant aspects of chemistry.

Chemistry is the science that deals with the properties, composition, and structure of substances (defined as elements and compounds), the transformations they undergo, and the energy that is released or absorbed during these processes. Every substance, whether naturally occurring or artificially produced, consists of one or more of the hundred-odd species of atoms that have been identified as elements. Although these atoms, in turn, are composed of more elementary particles, they are the basic building blocks of chemical substances; there is no quantity of oxygen, mercury, or gold, for example, smaller than an atom of that substance. Chemistry, therefore, is concerned not with the subatomic domain but with the properties of atoms and the laws governing their combinations and how the knowledge of these properties can be used to achieve specific purposes.

The great challenge in chemistry is the development of a coherent explanation of the complex behaviour of materials, why they appear as they do, what gives them their enduring properties, and how interactions among different substances can bring about the formation of new substances and the destruction of old ones. From the earliest attempts to understand the material world in rational terms, chemists have struggled to develop theories of matter that satisfactorily explain both permanence and change. The ordered assembly of indestructible atoms into small and large molecules, or extended networks of intermingled atoms, is generally accepted as the basis of permanence, while the reorganization of atoms or molecules into different arrangements lies behind theories of change. Thus chemistry involves the study of the atomic composition and structural architecture of substances, as well as the varied interactions among substances that can lead to sudden, often violent reactions.

Chemistry also is concerned with the utilization of natural substances and the creation of artificial ones. Cooking, fermentation, glass making, and metallurgy are all chemical processes that date from the beginnings of civilization. Today, vinyl, Teflon, liquid crystals, semiconductors, and superconductors represent the fruits of chemical technology. The 20th century has seen dramatic advances in the comprehension of the marvelous and complex chemistry of living organisms, and a molecular interpretation of health and disease holds great promise. Modern chemistry, aided by increasingly sophisticated instruments, studies materials as small as single atoms and as large and complex as DNA (deoxyribonucleic acid), which contains millions of atoms. New substances can even be designed to bear desired characteristics and then synthesized. The rate at which chemical knowledge continues to accumulate is remarkable. Over time more than 8,000,000 different chemical substances, both natural and artificial, have been characterized and produced. The number was less than 500,000 as recently as 1965.

Intimately interconnected with the intellectual challenges of chemistry are those associated with industry. In the mid-19th century the German chemist Justus von Liebig commented that the wealth of a nation could be gauged by the amount of sulfuric acid it produced. This acid, essential to many manufacturing processes, remains today the leading chemical product of industrialized countries. As Liebig recognized, a country that produces large amounts of sulfuric acid is one with a strong chemical industry and a strong economy as a whole. The production, distribution, and utilization of a wide range of chemical products is common to all highly developed nations. In fact, one can say that the "iron age" of civilization is being replaced by a "polymer age," for in some countries the total volume of polymers now produced exceeds that of iron.

Scope of Chemistry

The days are long past when one person could hope to have a detailed knowledge of all areas of chemistry. Those pursuing their interests into specific areas of chemistry communicate with others who share the same interests. Over time a group of chemists with specialized research interests become the founding members of an area of specialization. The areas of specialization that emerged early in the history of chemistry, such as organic, inorganic, physical, analytical, and industrial chemistry, along with biochemistry, remain of greatest general interest. There has been, however, much growth in the areas of polymer, environmental, and medicinal chemistry during the 20th century. Moreover, new specialities continue to appear, as, for example, pesticide, forensic, and computer chemistry.

Analytical Chemistry

Most of the materials that occur on Earth, such as wood, coal, minerals, or air, are mixtures of many different and distinct chemical substances. Each pure chemical substance (e.g., oxygen, iron, or water) has a characteristic set of properties that gives it its chemical identity. Iron, for example, is a common silver-white metal that melts at 1,535 °C, is very malleable, and readily combines with oxygen to form the common substances hematite and magnetite. The detection of iron in a mixture of metals, or in a compound such as magnetite, is a branch of analytical chemistry called qualitative analysis. Measurement of the actual amount of a certain substance in a compound or mixture is termed quantitative analysis. Quantitative analytic measurement has determined, for instance, that iron makes up 72.3 percent, by mass, of magnetite, the mineral commonly seen as black sand along beaches and stream banks. Over the years, chemists have discovered chemical reactions that indicate the presence of such elemental substances by the production of easily visible and identifiable products. Iron can be detected by chemical means if it is present in a sample to an amount of 1 part per million or greater. Some very simple qualitative tests reveal the presence of specific chemical elements in even smaller amounts. The yellow colour imparted to a flame by sodium is visible if the sample being ignited has as little as one-billionth of a gram of sodium. Such analytic tests have allowed chemists to identify the types and amounts of impurities in various substances and to determine the properties of very pure materials. Substances used in common laboratory experiments generally have impurity levels of less than 0.1 percent. For special applications, one can purchase chemicals that have impurities totaling less than 0.001 percent. The identification of pure substances and the analysis of chemical mixtures enable all other chemical disciplines to flourish.

The importance of analytical chemistry has never been greater than it is today. The demand in modern societies for a variety of safe foods, affordable consumer goods, abundant energy, and labour-saving technologies places a great burden on the environment. All chemical manufacturing produces waste products in addition to the desired substances, and waste disposal has not always been carried out carefully. Disruption of the environment has occurred since the dawn of civilization, and pollution problems have increased with the growth of global population. The techniques of analytical chemistry are relied on heavily to maintain a benign environment. The undesirable substances in water, air, soil, and food must be identified, their point of origin fixed, and safe, economical methods for their removal or neutralization developed. Once the amount of a pollutant deemed to be hazardous has been assessed, it becomes important to detect harmful substances at concentrations well below the danger level. Analytical chemists seek to develop increasingly accurate and sensitive techniques and instruments.

Sophisticated analytic instruments, often coupled with computers, have improved the accuracy with which chemists can identify substances and have lowered detection limits. An analytic technique in general use is gas chromatography, which separates the different components of a gaseous mixture by passing the mixture through a long, narrow column of absorbent but porous material. The different gases interact differently with this absorbent material and pass through the column at different rates. As the separate gases flow out of the column, they can be passed into another analytic instrument called a mass spectrometer, which separates substances according to the mass of their constituent ions. A combined gas chromatograph–mass spectrometer can rapidly identify the individual components of a chemical mixture whose concentrations may be no greater than a few parts per billion. Similar or even greater sensitivities can be obtained under favourable conditions using techniques such as atomic absorption, polarography, and neutron activation. The rate of instrumental innovation is such that analytic instruments often become obsolete within 10 years of their introduction. Newer instruments are more accurate and faster and are employed widely in the areas of environmental and medicinal chemistry.

Inorganic Chemistry

Modern chemistry, which dates more or less from the acceptance of the law of conservation of mass in the late 18th century, focused initially on those substances that were not associated with living organisms. Study of such substances, which normally have little or no carbon, constitutes the discipline of inorganic chemistry. Early work sought to identify the simple substances—namely, the elements—that are the constituents of all more complex substances. Some elements, such as gold and carbon, have been known since antiquity, and many others were discovered and studied throughout the 19th and early 20th centuries. Today, more than 100 are known. The study of such simple inorganic compounds as sodium chloride (common salt) has led to some of the fundamental concepts of modern chemistry, the law of definite proportions providing one notable example. This law states that for most pure chemical substances the constituent elements are always present in fixed proportions by mass (e.g., every 100 grams of salt contains 39.3 grams of sodium and 60.7 grams of chlorine). The crystalline form of salt, known as halite, consists of intermingled sodium and chlorine atoms, one sodium atom for each one of chlorine. Such a compound, formed solely by the combination of two elements, is known as a binary compound. Binary compounds are very common in inorganic chemistry, and they exhibit little structural variety. For this reason, the number of inorganic compounds is limited in spite of the large number of elements that may react with each other. If three or more elements are combined in a substance, the structural possibilities become greater.

After a period of quiescence in the early part of the 20th century, inorganic chemistry has again become an exciting area of research. Compounds of boron and hydrogen, known as boranes, have unique structural features that forced a change in thinking about the architecture of inorganic molecules. Some inorganic substances have structural features long believed to occur only in carbon compounds, and a few inorganic polymers have even been produced. Ceramics are materials composed of inorganic elements combined with oxygen. For centuries ceramic objects have been made by strongly heating a vessel formed from a paste of powdered minerals. Although ceramics are quite hard and stable at very high temperatures, they are usually brittle. Currently, new ceramics strong enough to be used as turbine blades in jet engines are being manufactured. There is hope that ceramics will one day replace steel in components of internal-combustion engines. In 1987 a ceramic containing yttrium, barium, copper, and oxygen, with the approximate formula YBa2Cu3O7, was found to be a superconductor at a temperature of about 100 K. A superconductor offers no resistance to the passage of an electrical current, and this new type of ceramic could very well find wide use in electrical and magnetic applications. A superconducting ceramic is so simple to make that it can be prepared in a high school laboratory. Its discovery illustrates the unpredictability of chemistry, for fundamental discoveries can still be made with simple equipment and inexpensive materials.

Many of the most interesting developments in inorganic chemistry bridge the gap with other disciplines. Organometallic chemistry investigates compounds that contain inorganic elements combined with carbon-rich units. Many organometallic compounds play an important role in industrial chemistry as catalysts, which are substances that are able to accelerate the rate of a reaction even when present in only very small amounts. Some success has been achieved in the use of such catalysts for converting natural gas to related but more useful chemical substances. Chemists also have created large inorganic molecules that contain a core of metal atoms, such as platinum, surrounded by a shell of different chemical units. Some of these compounds, referred to as metal clusters, have characteristics of metals, while others react in ways similar to biologic systems. Trace amounts of metals in biologic systems are essential for processes such as respiration, nerve function, and cell metabolism. Processes of this kind form the object of study of bioinorganic chemistry. Although organic molecules were once thought to be the distinguishing chemical feature of living creatures, it is now known that inorganic chemistry plays a vital role as well.

Organic Chemistry

Organic compounds are based on the chemistry of carbon. Carbon is unique in the variety and extent of structures that can result from the three-dimensional connections of its atoms. The process of photosynthesis converts carbon dioxide and water to oxygen and compounds known as carbohydrates. Both cellulose, the substance that gives structural rigidity to plants, and starch, the energy storage product of plants, are polymeric carbohydrates. Simple carbohydrates produced by photosynthesis form the raw material for the myriad organic compounds found in the plant and animal kingdoms. When combined with variable amounts of hydrogen, oxygen, nitrogen, sulfur, phosphorus, and other elements, the structural possibilities of carbon compounds become limitless, and their number far exceeds the total of all nonorganic compounds. A major focus of organic chemistry is the isolation, purification, and structural study of these naturally occurring substances. Many natural products are simple molecules. Examples include formic acid (HCO_2H) in ants, ethyl alcohol (C_2H_5OH) in fermenting fruit, and oxalic acid ($C_2H_2O_4$) in rhubarb leaves. Other natural products, such as penicillin, vitamin B12, proteins, and nucleic acids, are exceedingly complex. The isolation of pure natural products from their host organism is made difficult by the low concentrations in which they may be present. Once they are isolated in pure form, however, modern instrumental techniques can reveal structural details for amounts weighing as little as one-millionth of a gram. The correlation of the physical and chemical properties of compounds with their structural features is the domain of physical organic chemistry. Once the properties endowed upon a substance by specific structural units termed functional groups are known, it becomes possible to design novel molecules that may exhibit desired properties. The preparation, under controlled laboratory conditions, of specific compounds is known as synthetic chemistry. Some products are easier to synthesize than to collect and purify from their natural sources. Tons of vitamin C, for example, are synthesized annually. Many synthetic substances have novel properties that make them especially useful. Plastics are a prime example, as are many drugs and agricultural chemicals. A continuing challenge for synthetic chemists is the structural complexity of most organic substances. To synthesize a desired substance, the atoms must be pieced together in the correct order and with the proper three-dimensional relationships. Just as a given pile of lumber and bricks can be assembled in many ways to build houses of several different designs, so too can a fixed number of atoms be connected together in various ways to give different molecules. Only one structural arrangement out of the many possibilities will be identical with a naturally occurring molecule. The antibiotic erythromycin, for example, contains 37 carbon, 67 hydrogen, and 13 oxygen atoms, along with one nitrogen atom. Even when joined together in the proper order, these 118 atoms can give rise to 262,144 different structures, only one of which has the characteristics of natural erythromycin. The great abundance of organic compounds, their fundamental role in the chemistry of life, and their structural diversity have made their study especially challenging and exciting. Organic chemistry is the largest area of specialization among the various fields of chemistry.

Biochemistry

As understanding of inanimate chemistry grew during the 19th century, attempts to interpret the physiological processes of living organisms in terms of molecular structure and reactivity gave rise to the discipline of biochemistry. Biochemists employ the techniques and theories of chemistry to probe the molecular basis of life. An organism is investigated on the premise that its physiological processes are the consequence of many thousands of chemical reactions occurring in a highly integrated manner. Biochemists have established, among other things, the principles that underlie energy transfer in cells, the chemical structure of cell membranes, the coding and transmission of hereditary information, muscular and nerve function, and biosynthetic pathways. In fact, related biomolecules have been found to fulfill similar roles in organisms as different as bacteria and human beings. The study of biomolecules, however, presents many difficulties. Such molecules are often very large and exhibit great structural complexity; moreover, the chemical reactions they undergo are usually exceedingly fast. The separation of the two strands of DNA, for instance, occurs in one-millionth of a second. Such rapid rates of reaction are possible only through the intermediary action of biomolecules called enzymes. Enzymes are proteins that owe their remarkable rate-accelerating abilities to their three-dimensional chemical structure. Not surprisingly, biochemical discoveries have had a great impact on the understanding and treatment of disease. Many ailments due to inborn errors of metabolism have been traced to specific genetic defects. Other diseases result from disruptions in normal biochemical pathways.

Frequently, symptoms can be alleviated by drugs, and the discovery, mode of action, and degradation of therapeutic agents is another of the major areas of study in biochemistry. Bacterial infections can be treated with sulfonamides, penicillins, tetracyclines and research into viral infections has revealed the effectiveness of acyclovir against the herpes virus. There is much current interest in the details of carcinogenesis and cancer chemotherapy. It is known, for example, that cancer can result when cancer-causing molecules, or carcinogens as they are called, react with nucleic acids and proteins and interfere with their normal modes of action. Researchers have developed tests that can identify molecules likelyto be carcinogenic. The hope, of course, is that progress in the prevention and treatment of cancer will accelerate once the biochemical basis of the disease is more fully understood.

The molecular basis of biologic processes is an essential feature of the fast-growing disciplines of molecular biology and biotechnology. Chemistry has developed methods for rapidly and accurately determining the structure of proteins and DNA. In addition, efficient laboratory methods for the synthesis of genes are being devised. Ultimately, the correction of genetic diseases by replacement of defective genes with normal ones may become possible.

Polymer Chemistry

The simple substance ethylene is a gas composed of molecules with the formula CH_2CH_2. Under certain conditions, many ethylene molecules will join together to form a long chain called polyethylene, with the formula $(CH_2CH_2)_n$, where n is a variable but large number. Polyethylene is a tough, durable solid material quite different from ethylene. It is an example of a polymer, which is a large molecule made up of many smaller molecules (monomers), usually joined together in a linear fashion. Many naturally occurring substances, including cellulose, starch, cotton, wool, rubber, leather, proteins, and DNA, are polymers. Polyethylene, nylon, and acrylics are examples of synthetic polymers. The study of such materials lies within the domain of polymer chemistry, a specialty that has flourished in the 20th century. The investigation of natural polymers overlaps considerably with biochemistry, but the synthesis of new polymers, the investigation of polymerization processes, and the characterization of the structure and properties of polymeric materials all pose unique problems for polymer chemists.

Polymer chemists have designed and synthesized polymers that vary in hardness, flexibility, softening temperature, solubility in water, and biodegradability. They have produced polymeric materials that are as strong as steel yet lighter and more resistant to corrosion. Oil, natural gas, and water pipelines are now routinely constructed of plastic pipe. In recent years, automakers have increased their use of plastic components to build lighter vehicles that consume less fuel. Other industries such as those involved in the manufacture of textiles, rubber, paper, and packaging materials are built upon polymer chemistry.

Besides producing new kinds of polymeric materials, researchers are concerned with developing special catalysts that are required by the large-scale industrial synthesis of commercial polymers. Without such catalysts, the polymerization process would be very slow in certain cases.

Physical Chemistry

Many chemical disciplines, such as those already discussed, focus on certain classes of materials that share common structural and chemical features. Other specialties may be centred not on a class of substances but rather on their interactions and transformations. The oldest of these fields is physical chemistry, which seeks to measure, correlate, and explain the quantitative aspects of chemical processes. The Anglo-Irish chemist Robert Boyle, for example, discovered in the 17th century that at room temperature the volume of a fixed quantity of gas decreases proportionally as the pressure on it increases. Thus, for a gas at constant temperature, the product of its volume V and pressure P equals a constant number—i.e., PV = constant. Such a simple arithmetic relationship is valid for nearly all gases at room temperature and at pressures equal to or less than one atmosphere. Subsequent work has shown that the relationship loses its validity at higher pressures, but more complicated expressions that more accurately match experimental results can be derived. The discovery and investigation of such chemical regularities, often called laws of nature, lie within the realm of physical chemistry. For much of the 18th century the source of mathematical regularity in chemical systems was assumed to be the continuum of forces and fields that surround the atoms making up chemical elements and compounds.

Developments in the 20th century, however, have shown that chemical behaviour is best interpreted by a quantum mechanical model of atomic and molecular structure. The branch of physical chemistry that is largely devoted to this subject is theoretical chemistry. Theoretical chemists make extensive use of computers to help them solve complicated mathematical equations. Other branches of physical chemistry include chemical thermodynamics, which deals with the relationship between heat and other forms of chemical energy, and chemical kinetics, which seeks to measure and understand the rates of chemical reactions. Electrochemistry investigates the interrelationship of electric current and chemical change. The passage of an electric current through a chemical solution causes changes in the constituent substances that are often reversible—i.e., under different conditions the altered substances themselves will yield an electric current. Common batteries contain chemical substances that, when placed in contact with each other by closing an electrical circuit, will deliver current at a constant voltage until the substances are consumed. At present there is much interest in devices that can use the energy in sunlight to drive chemical reactions whose products are capable of storing the energy. The discovery of such devices would make possible the widespread utilization of solar energy.

There are many other disciplines within physical chemistry that are concerned more with the general properties of substances and the interactions among substances than with the substances themselves. Photochemistry is a specialty that investigates the interaction of light with matter. Chemical reactions initiated by the absorption of light can be very different from those that occur by other means. Vitamin D, for example, is formed in the human body when the steroid ergosterol absorbs solar radiation; ergosterol does not change to vitamin D in the dark.

A rapidly developing sub-discipline of physical chemistry is surface chemistry. It examines the properties of chemical surfaces, relying heavily on instruments that can provide a chemical profile of such surfaces. Whenever a solid is exposed to a liquid or a gas, a reaction occurs initially on the surface of the solid, and its properties can change dramatically as a result. Aluminum is a case in point: it is resistant to corrosion precisely because the surface of the pure metal reacts with oxygen to form a layer of aluminum oxide, which serves to protect the interior of the metal from further oxidation. Numerous reaction catalysts perform their function by providing a reactive surface on which substances can react.

Industrial Chemistry

The manufacture, sale, and distribution of chemical products is one of the cornerstones of a developed country. Chemists play an important role in the manufacture, inspection, and safe handling of chemical products, as well as in product development and general management. The manufacture of basic chemicals such as oxygen, chlorine, ammonia, and sulfuric acid provides the raw materials for industries producing textiles, agricultural products, metals, paints, and pulp and paper. Specialty chemicals are produced in smaller amounts for industries involved with such products as pharmaceuticals, foodstuffs, packaging, detergents, flavours, and fragrances. To a large extent, the chemical industry takes the products and reactions common to "bench-top" chemical processes and scales them up to industrial quantities.

The monitoring and control of bulk chemical processes, especially with regard to heat transfer, pose problems usually tackled by chemists and chemical engineers. The disposal of by-products also is a major problem for bulk chemical producers. These and other challenges of industrial chemistry set it apart from the more purely intellectual disciplines of chemistry discussed above. Yet, within the chemical industry, there is a considerable amount of fundamental research undertaken within traditional specialties. Most large chemical companies have research-and-development capability. Pharmaceutical firms, for example, operate large research laboratories in which chemists test molecules for pharmacological activity. The new products and processes that are discovered in such laboratories are often patented and become a source of profit for the company funding the research. A great deal of the research conducted in the chemical industry can be termed applied research because its goals are closely tied to the products and processes of the company concerned. New technologies often require much chemical expertise. The fabrication of, say, electronic microcircuits involves close to 100 separate chemical steps from start to finish. Thus, the chemical industry evolves with the technological advances of the modern world and at the same time often contributes to the rate of progress.

Chemical Elements and Compounds

Elements

A chemical element is a pure substance that consists of one type of atom. Each atom has an atomic number, which represents the number of protons that are in the nucleus of a single atom of that element. The periodic table of elements is ordered by ascending atomic number.

The chemical elements are divided into the metals, the metalloids, and the non-metals. Metals, typically found on the left side of the periodic table, are:

- Often conductive to electricity;
- Malleable;
- Shiny;
- Sometimes magnetic.

Aluminum, iron, copper, gold, mercury and lead are metals.

In contrast, non-metals, found on the right side of the periodic table (to the right of the staircase), are:

- Typically not conductive;
- Not malleable;
- Dull (not shiny);
- Not magnetic.

Examples of elemental non-metals include carbon and oxygen.

Metalloids have some characteristics of metals and some characteristics of non-metals. Silicon and arsenic are metalloids.

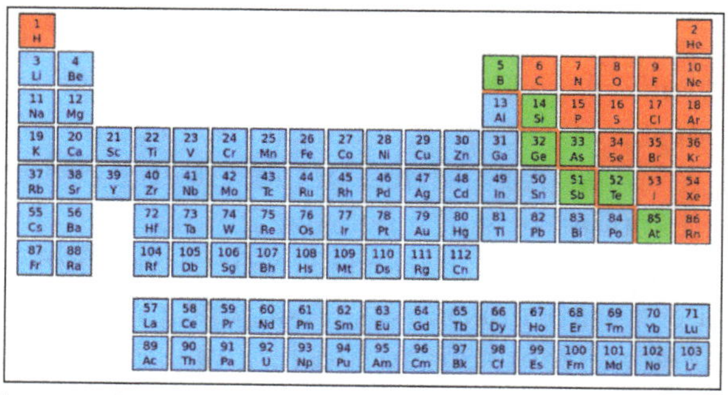

The periodic table: The periodic table shows 118 elements, including metals (blue), nonmetals (red), and metalloids (green).

As of November, 2011, 118 elements have been identified. Of these 118 known elements, only the first 98 are known to occur naturally on Earth. The elements that do not occur naturally on Earth are the synthetic products of man-made nuclear reactions. 80 of the 98 naturally-occurring elements are stable; the rest are radioactive, which means they decay into lighter elements over timescales ranging from fractions of a second to billions of years.

Hydrogen and helium are by far the most abundant elements in the universe. However, iron is the most abundant element (by mass) in the composition of the Earth, and oxygen is the most common element in the layer that is the Earth's crust.

Although all known chemical matter is composed of these elements, chemical matter itself constitutes only about 15% of the matter in the universe. The remainder is dark matter, a mysterious substance that is not composed of chemical elements. Dark matter lacks protons, neutrons, or electrons.

Compounds

Pure samples of isolated elements are uncommon in nature. While the 98 naturally occurring elements have all been identified in mineral samples from the Earth's crust, only a small minority of them can be found as recognizable, relatively pure minerals. Among the more common of such "native elements" are copper, silver, gold, and sulfur. Carbon is also commonly found in the form of coal, graphite, and diamonds. The noble gases (e.g., neon) and noble metals (e.g., mercury) can also be found in their pure, non-bonded forms in nature. Still, most of these elements are found in mixtures.

When two distinct elements are chemically combined—i.e., chemical bonds form between their atoms—the result is called a chemical compound. Most elements on Earth bond with other elements to form chemical compounds, such as sodium (Na) and Chloride (Cl), which combine to form table salt (NaCl). Water is another example of a chemical compound. The two or more component elements of a compound can be separated through chemical reactions.

Chemical compounds have a unique and defined structure, which consists of a fixed ratio of atoms held together in a defined spatial arrangement by chemical bonds. Chemical compounds can be:

- Molecular compounds held together by covalent bonds;
- Salts held together by ionic bonds;
- Intermetallic compounds held together by metallic bonds;
- Complexes held together by coordinate covalent bonds.

Pure chemical elements are not considered chemical compounds, even if they consist of diatomic or polyatomic molecules (molecules that contain only multiple atoms of a single element, such as H_2 or S_8).

Classification of Compounds

Chemical compounds may be classified according to several different criteria. One common method is based on the specific elements present. For example, oxides contain one or more oxygen atoms, hydrides contain one or more hydrogen atoms, and halides contain one or more halogen (Group 17) atoms. Organic compounds are characterized as those compounds with a backbone of carbon atoms, and all the remaining compounds are classified as inorganic. As the name suggests, organometallic compounds are organic compounds bonded to metal atoms.

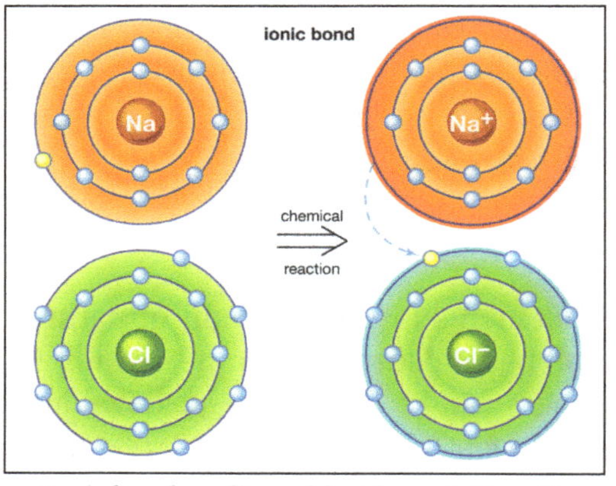

Ionic bond: sodium chloride, or table salt.

Ionic bonding in sodium chloride: An atom of sodium (Na) donates one of its electrons to an atom of chlorine (Cl) in a chemical reaction, and the resulting positive ion (Na^+) and negative ion (Cl^-) form a stable ionic compound (sodium chloride; common table salt) based on this ionic bond.

Another classification scheme for chemical compounds is based on the types of bonds that the compound contains. Ionic compounds contain ions and are held together by the attractive forces among the oppositely charged ions. Common salt (sodium chloride) is one of the best-known ionic compounds. Molecular compounds contain discrete molecules, which are held together by sharing electrons (covalent bonding). Examples are water, which contains H_2O molecules; methane, which contains CH_4 molecules; and hydrogen fluoride, which contains HF molecules.

A third classification scheme is based on reactivity—specifically, the types of chemical reactions that the compounds are likely to undergo. For example, acids are compounds that produce H^+ ions (protons) when dissolved in water to produce aqueous solutions. Thus, acids are defined as proton donors. The most common acids are aqueous solutions of HCl (hydrochloric acid), H_2SO_4 (sulfuric acid), HNO_3 (nitric acid), and H_3PO_4 (phosphoric acid). Bases, on the other hand, are proton acceptors. The most common base is the hydroxide ion(OH^-), which reacts with an H^+ ion to form a water molecule.

$$H^+ + OH^- \rightarrow HOH \text{ (usually written } H_2O)$$

Oxidation-reduction reactions constitute another important class of chemical reactions. Oxidation involves a loss of electrons, whereas reduction involves a gain of electrons. For example, in the reaction between sodium metal and chlorine gas to form sodium chloride,

$$2Na + Cl_2 \rightarrow 2NaCl$$

Electrons (e^-) are transferred from sodium atoms to chlorine atoms to form Na^+ and Cl^- ions in the reaction product, sodium chloride.

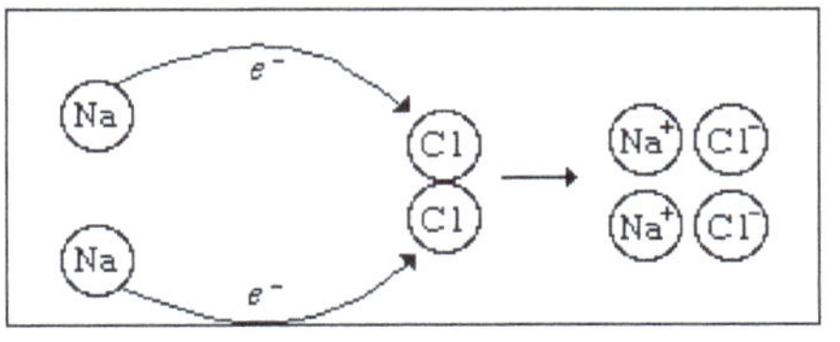

In this process, each sodium atom loses an electron and is thus oxidized, and each chlorine atom gains an electron and is thus reduced. In this reaction, sodium is called the reducing agent (it furnishes electrons), and chlorine is called the oxidizing agent (it consumes electrons). The most common reducing agents are metals, for they tend to lose electrons in their reactions with nonmetals. The most common oxidizing agents are halogens—such as fluorine (F_2), chlorine (Cl_2), and bromine (Br_2)—and certain oxy anions, such as the permanganate ion (MnO_4^-) and the dichromate ion ($Cr_2O_7^{2-}$).

Inorganic Compounds

Inorganic compounds include compounds that are made up of two or more elements other than carbon, as well as certain carbon-containing compounds that lack carbon-carbon bonds, such as cyanides and carbonates. Inorganic compounds are most often classified in terms of the elements or groups of elements that they contain. Oxides, for example, can be either ionic or molecular. Ionic oxides contain O^{2-} (oxide) ions and metal cations, whereas molecular oxides contain molecules in which oxygen (O) is covalently bonded to other nonmetals such as sulfur(S) or nitrogen (N). When ionic oxides are dissolved in water, the O^{2-} ions react with water molecules to form hydroxide ions (OH^-), and a basic solution results. Molecular oxides react with water to produce oxyacids, such as sulfuric acid (H_2SO_4) and nitric acid (HNO_3). In addition, inorganic compounds include hydrides (containing hydrogen atoms or H^- ions), nitrides (containing N^{3-} ions), phosphides (containing P^{3-} ions), and sulfides (containing S^{2-}ions).

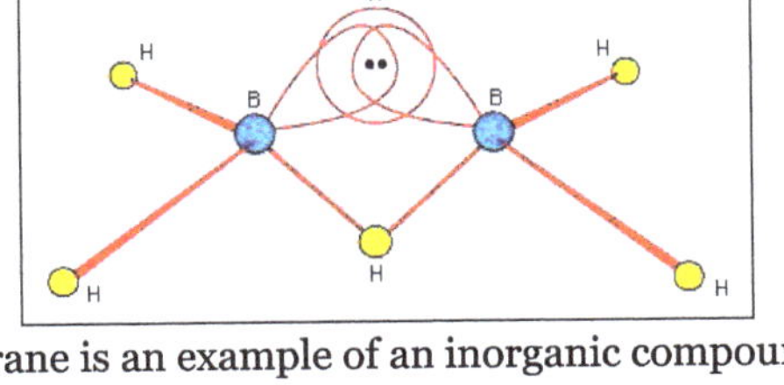

Borane is an example of an inorganic compound.

Transition metals form a great variety of inorganic compounds. The most important of these are coordination compounds in which the metal atom or ion is surrounded by two to six ligands. Ligands are ions or neutral molecules with electron pairs that they can donate to the metal atom to form a coordinate-covalent bond.

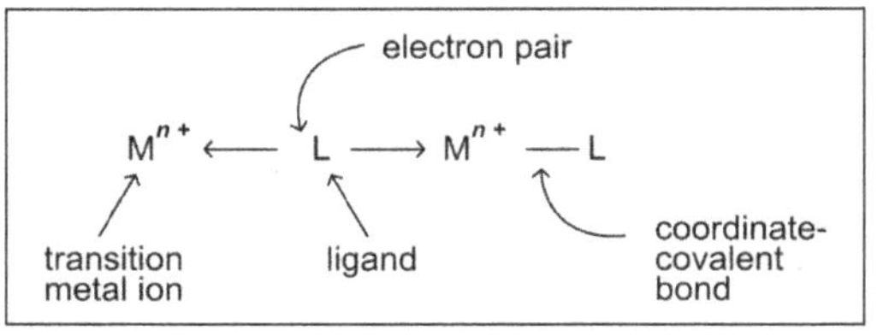

The resulting covalent bond is given a special name because one entity (the ligand) furnishes both of the electrons that are subsequently shared in the bond. An example of a coordination compound is $[Co(NH_3)_6]Cl_3$, which contains the $Co(NH_3)_6^{3+}$ ion, a cobalt ion (Co^{3+}) with six ammonia molecules (NH_3) attached to it, acting as ligands.

In the early days of the science of chemistry, there was no systematic approach to naming compounds. Chemists coined names such as sugar of lead, quicklime, milk of magnesia, Epsom salts, and laughing gas to describe familiar compounds. Such names are called common or trivial names. As chemistry advanced, it became evident that, if common names were used for all known compounds, which number in the millions, great confusion would result. It clearly would be impossible to memorize trivial names for such a large number of compounds. Therefore a systematic nomenclature (naming process) has been developed. There are, however, certain familiar compounds that are always referred to by their common names. The systematic names for H_2O and NH_3, for example, are never used; these vital compounds are known only as water and ammonia, respectively.

The simplest chemical compounds are binary compounds—those consisting of two elements. Different rules apply for the nomenclature of binary ionic compounds and binary molecular (covalent) compounds, and so they will be considered separately.

Binary Compounds

Binary Ionic Compounds

The nomenclature for binary ionic compounds simply entails naming the ions according to the following rules:

1. The positive ion (called a cation) is named first and the negative ion (anion) second.

2. A simple cation (obtained from a single atom) takes its name from its parent element. For example, Li^+ is called lithium in the names of compounds containing this ion. Similarly, Na^+ is called sodium, Mg^{2+} is called magnesium, and so on.

3. A simple anion (obtained from a single atom) is named by taking the root of the parent element's name and adding the suffix -ide. Thus, the F^- ion is called fluoride, Br^- is called bromide, S^{2-} is called sulfide, and so on.

The following examples illustrate the nomenclature rules for binary ionic compounds:

compound	ions present	name
NaCl	Na^+, Cl^-	sodium chloride
KI	K^+, I^-	potassium iodide
CaS	Ca^{2+}, S^{2-}	calcium sulfide
CsBr	Cs^+, Br^-	cesium bromide
MgO	Mg^{2-}, O^{2-}	magnesium oxide

In the formulas of ionic compounds, simple ions are represented by the chemical symbol for the element: Cl means Cl^-, Na means Na^+, and so on. When individual ions are shown, however, the charge is always included. Thus, the formula of potassium bromide is given as KBr, but, when the potassium and bromide ions are shown individually, they are written K^+ and Br^-.

When a given metal atom can form more than one type of cation, the charge on the particular cation present must be specified in the name of the compound. For example, lead (Pb) can exist as Pb^{2+} or Pb^{4+} ions in ionic compounds. Also, iron (Fe) can form Fe^{2+} or Fe^{3+} ions, tin (Sn) can form Sn^{2+} or Sn^{4+} ions, gold (Au) can form Au^+ or Au^{3+} ions, and so on. Therefore, the names of binary compounds containing metals such as these must include a Roman numeralto specify the charge on the ion. For example, the compound $FeCl_3$, which contains Fe^{3+}, is named iron(III) chloride. On the other hand, the compound $FeCl_2$, which contains Fe^{2+}, is designated as iron(II) chloride. In each case, the Roman numeral in the name specifies the charge of the metal ion present.

Common simple cations and anions			
cation	name	anion	name
H^+	hydrogen	H^-	hydride
Li^+	lithium	F^-	fluoride
Na^+	sodium	Cl^-	chloride
K^+	potassium	Br^-	bromide
Cs^+	cesium	I^-	iodide
Be^{2+}	beryllium	O^{2-}	oxide
Mg^{2+}	magnesium	S^{2-}	sulfide
Ca^{2+}	calcium		
Ba^{2+}	barium		
Al^{3+}	aluminum		
Ag^+	silver		

An alternative system for naming compounds containing metals that form only two ions is sometimes seen, especially in older literature. The ion with the higher charge has a name ending in -ic, and the one with the lower charge has the suffix ous. For example, Fe^{3+} is called the ferric ion, and Fe^{2+} is called the ferrous ion. The names for $FeCl_3$ and $FeCl_2$ are then ferric chloride and ferrous chloride, respectively.

Common ions that form multiple cations		
ion	systematic name	alternate name
Fe^{3+}	iron(III)	ferric
Fe^{2+}	iron(II)	ferrous
Cu^{2+}	copper(II)	cupric
Cu^{+}	copper(I)	cuprous
Co^{3+}	cobalt(III)	cobaltic
Co^{2+}	cobalt(II)	cobaltous
Sn^{4+}	tin(IV)	stannic
Sn^{2+}	tin(II)	stannous
Pb^{4+}	lead(IV)	plumbic
Pb^{2+}	lead(II)	plumbous
Hg^{2+}	mercury(II)	mercuric
Hg_2^{2+}	mercury(I)	mercurous

Binary Molecular (Covalent) Compounds

Binary molecular (covalent) compounds are formed as the result of a reaction between two nonmetals. Although there are no ions in these compounds, they are named in a similar manner to binary ionic compounds. The nomenclature of binary covalent compounds follows these rules:

1. The first element in the formula is given first, using the element's full name.
2. The second element is named as if it were an anion.
3. Prefixes are used to denote the numbers of atoms present. If the first element exists as a single atom, the prefix mono- is omitted. For example, CO is called carbon monoxide rather than monocarbon monoxide.

These examples show how the rules are applied for the covalent compounds formed by nitrogen and oxygen:

compound	systematic name	common name
N_2O	dinitrogen monoxide	nitrous oxide (laughing gas)
NO	nitrogen monoxide	nitric oxide
NO_2	nitrogen dioxide	
N_2O_3	dinitrogen trioxide	

compound	systematic name	common name
N_2O_4	dinitrogen tetroxide	
N_2O_5	dinitrogen pentoxide	

To avoid awkward pronunciations, the final o or a of the prefix is often dropped when the element name begins with a vowel. For example, N_2O_4 is referred to as dinitrogen tetroxide, not dinitrogen tetraoxide, and CO is called carbon monoxide, not carbon monooxide.

Prefixes used in chemical nomenclature	
prefix	number of atoms
mono-	1
di-	2
tri-	3
tetra-	4
penta-	5
hexa-	6
hepta-	7
octa-	8

Nonbinary Compounds

Ionic Compounds Containing Polyatomic Ions

A special type of ionic compound is exemplified by ammonium nitrate (NH_4NO_3), which contains two polyatomic ions, NH_4^+ and NO_3^-. As the name suggests, a polyatomic ion is a charged entity composed of several atoms bound together. Polyatomic ions have special names that are used in the nomenclature of the compounds containing them.

Common polyatomic ions			
ion	name	ion	name
NH_4^+	ammonium	CO_3^{2-}	carbonate
NO_2^-	nitrite	HCO_3^-	hydrogen carbonate
NO_3^-	nitrate	ClO^-	hypochlorite
SO_3^{2-}	sulfite	ClO_2^-	chlorite
SO_4^{2-}	sulfate	ClO_3^-	chlorate
HSO_4^-	hydrogen sulfate	ClO_4^-	perchlorate
OH^-	hydroxide	$C_2H_3O_2^-$	acetate
CN^-	cyanide	MnO_4^-	permanganate
PO_4^{3-}	phosphate	$Cr_2O_7^{2-}$	dichromate
HPO_4^{2-}	hydrogen phosphate	CrO_4^{2-}	chromate
$H_2PO_4^-$	dihydrogen phosphate	O_2^{2-}	peroxide

Bisulfate and bicarbonate are widely used common names for hydrogen sulfate and hydrogen carbonate, respectively.

Several series of polyatomic anions exist that contain an atom of a given element in combination with different numbers of oxygen atoms. Such anions are called oxy anions. When the series contains only two members, the name of the ion with fewer oxygen atoms ends in ite, and the name of the other ion ends in -ate. For example, SO_3^{2-} is called sulfite and SO_4^{2-} is called sulfate. In those cases where more than two oxy anions constitute the series, hypo- (less than) and per- (more than) are used as prefixes to name the members of the series with the smallest and the largest number of oxygen atoms, respectively.

The chlorine-containing oxy anions provide an example:

ClO^-	hypochlorite
ClO_2^-	chlorite
ClO_3^-	chlorate
ClO_4^-	perchlorate

Naming ionic compounds that contain polyatomic ions is similar to naming binary ionic compounds. For example, the compound NaOH is called sodium hydroxide, because it contains the Na^+ (sodium) cation and the OH^- (hydroxide) anion. As in binary ionic compounds, when a metal that can form multiple cations is present, a Roman numeral is required to specify the charge on the cation. For example, the compound $FeSO_4$ is called iron(II) sulfate, because it contains Fe^{2+}.

Acids

An acid can be thought of as a molecule containing at least one hydrogen cation (H^+) attached to an anion. The nomenclature of acids depends on whether the anion contains oxygen. If the anion does not contain oxygen, the acid is named with the prefix hydro- and the suffix -ic. For example, HCl dissolved in water is called hydrochloric acid. Likewise, HCN and H_{2S} dissolved in water are called hydrocyanic and hydrosulfuric acids, respectively.

If the anion of the acid contains oxygen, the name is formed by adding the suffix -ic or -ous to the root name of the anion. If the anion name ends in -ate, the -ate is replaced by -ic (or sometimes -ric). For example, H_2SO_4 contains the sulfate anion (SO_4^{2-}) and is called sulfuric acid; H_3PO_4 contains the phosphate anion (PO_4^{3-}) and is called phosphoric acid; and $HC_2H_3O_2$, which contains the acetate ion ($C_2H_3O_2^-$), is called acetic acid. For anions with an -ite ending, the -ite is replaced by -ous in naming the acid. For example, H_2SO_3, which contains sulfite (SO_3^{2-}), is called sulfurous acid; and HNO_2, which contains nitrite (NO_2^-), is named nitrous acid. The acids of the oxy anions of chlorine are used here to illustrate the rules for naming acids with oxygen-containing cations.

Names of less common acids	
formula	name
H_3BO_3	orthoboric acid
H_2CO_3	carbonic acid
H_3PO_4	orthophosphoric acid
$H_4P_2O_7$	pyrophosphoric acid
$H_5P_3O_{10}$	triphosphoric acid
$(HPO_3)n$	metaphosphoric acid
$(HPO_3)_3$	trimetaphosphoric acid
H_3PO_3	phosphorous acid
H_3PO_2	hypophosphorous acid
H_2SO_5	peroxosulfuric acid
$H_2S_2O_6$	dithionic acid
$H_2S_2O_3$	thiosulfuric acid
$HMnO_4$	permanganic acid

Often called boric acid. Often called phosphoric acid.

Names of common acids	
formula	name
HF	hydrofluoric acid
HCl	hydrochloric acid
HBr	hydrobromic acid
HI	hydroiodic acid
HCN	hydrocyanic acid
H_2S	hydrosulfuric acid
HNO_3	nitric acid
HNO_2	nitrous acid
H_2SO_4	sulfuric acid
H_2SO_3	sulfurous acid
$HC_2H_3O_2$	acetic acid

acid formula	anion	name
$HClO_4$	perchlorate	perchloric acid
$HClO_3$	chlorate	chloric acid
$HClO_2$	chlorite	chlorous acid
HClO	hypochlorite	hypochlorous acid

Compounds with Complex Ions

A coordination compound is composed of one or more complex structural units, each of which has a central atom bound directly to a surrounding set of groups called ligands. The nomenclature of coordination compounds is based on these structural relationships.

Organic Compounds

In general, organic compounds are substances that contain carbon (C), and carbon atoms provide the key structural framework that generates the vast diversity of organic compounds. All things on the Earth (and most likely elsewhere in the universe) that can be described as living have a crucial dependence on organic compounds. Foodstuffs—namely, fats, proteins, and carbohydrates—are organic compounds, as are such vital substances as hemoglobin, chlorophyll, enzymes, hormones, and vitamins. Other materials that add to the comfort, health, or convenience of humans are composed of organic compounds, including clothing made of cotton, wool, silk, and synthetic fibres; common fuels, such as wood, coal, petroleum, and natural gas; components of protective coatings, such as varnishes, paints, lacquers, and enamels; antibiotics and synthetic drugs; natural and synthetic rubber; dyes; plastics; and pesticides.

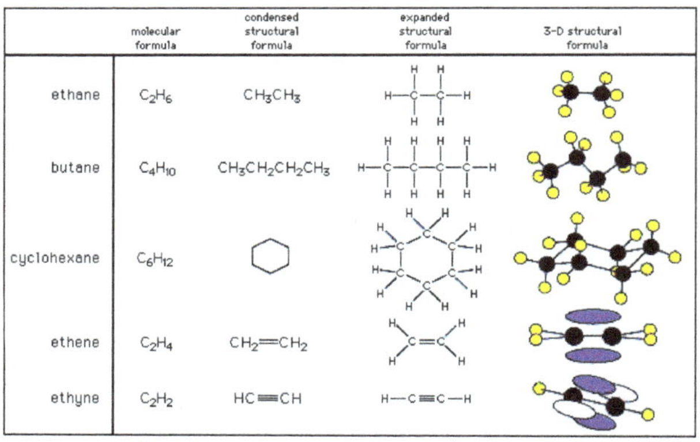

The structures of organic compounds can be depicted in condensed, expanded, and three-dimensional structural formulas.

Carbon Bonding

The carbon atom is unique among elements in its tendency to form extensive networks of covalent bonds not only with other elements but also with itself. Because of its position midway in the second horizontal row of the periodic table, carbon is neither an electropositive nor an electronegative element; it therefore is more likely to share electrons than to gain or lose them. Moreover, of all the elements in the second row, carbon has the maximum number of outer shell electrons (four) capable of forming covalent bonds. (Other elements, such as phosphorus [P] and cobalt [Co], are able to form five and six covalent bonds, respectively, with other elements, but they lack carbon's ability to bond indefinitely with itself.) When fully bonded to other atoms, the four bonds of the carbon atom are directed to the corners of a tetrahedron and make angles of about 109.5° with each other. The result is that not only can carbon atoms combine with one another indefinitely to give compounds of extremely high molecular weight, but the molecules formed can exist in an infinite variety of three-dimensional structures. The possibilities for diversity are increased by the presence of atoms other than carbon in organic compounds,especially hydrogen (H), oxygen (O), nitrogen (N), halogens (fluorine [F], chlorine [Cl], bromine [Br], and iodine [I]), and sulfur (S). It is the enormous potential for variation in chemical properties that has made organic compounds essential to life on Earth.

The structures of organic compounds commonly are represented by simplified structural formulas, which show not only the kinds and numbers of atoms present in the molecule but also the way in which the atoms are linked by the covalent bonds—information that is not given by simple molecular formulas, which specify only the number and type of atoms contained in a molecule. (With most inorganic compounds, the use of structural formulas is not necessary, because only a few atoms are involved and only a single arrangement of the atoms is possible.) In the structural formulas of organic compounds, short lines are used to represent the covalent bonds. Atoms of the individual elements are represented by their chemical symbols, as in molecular formulas.

Structural formulas vary widely in the amount of three-dimensional information they convey, and the type of structural formula used for any one molecule depends on the nature of the information the formula is meant to display. The different levels of sophistication can be illustrated by considering some of the least complex organic compounds, the hydrocarbons. The gas ethane, for example, has the molecular formula C_2H_6. The simplest structural formula, drawn either in a condensed or in an expanded version, reveals that ethane consists of two carbon atoms bonded to one another, each carbon atom bearing three hydrogen atoms. Such a two-dimensional representation correctly shows the bonding arrangement in ethane, but it does not convey any information about its three-dimensional architecture. A more sophisticated structural formula can be drawn to better represent the three-dimensional structure of the molecule. Such a structural formula correctly shows the tetrahedral orientation of the four atoms (one carbon and three hydrogens) bonded to each carbon, and the specific architecture of the molecule.

Larger organic molecules are formed by the addition of more carbon atoms. Butane, for example, is a gaseous hydrocarbon with the molecular formula C_4H_{10}, and it exists as a chain of four carbon atoms with 10 attached hydrogen atoms. As carbon atoms are added to a molecular framework, the carbon chain can develop branches or form cyclic structures. A very common ring structure contains six carbon atoms in a ring, each bonded in a tetrahedral arrangement, as in the hydrocarbon cyclohexane, C_6H_{12}. Such ring structures are often very simply represented as regular polygons in which each apex represents a carbon atom, and the hydrogen atoms that complete the bonding requirements of the carbon atoms are not shown. The polygon convention for cyclic structures reveals concisely the bonding arrangement of the molecule but does not explicitly convey information about the actual three-dimensional architecture. It should be noted that the polygon is only a two-dimensional symbol for the three-dimensional molecule.

Under certain bonding conditions, adjacent atoms will form multiple bonds with each other. A double bond is formed when two atoms use two electron pairs to form two covalent bonds; a triple bond results when two atoms share three electron pairs to form three covalent bonds. Multiple bonds have special structural and electronic features that generate interesting chemical properties. The six atoms involved in a double bond (as in ethene, C_2H_4) lie in a single plane, with regions above and below the plane occupied by the electrons of the second covalent bond. Atoms in a triple bond (as in acetylene, or ethyne, C_2H_2) lie in a straight line, with four regions beside the bond axis occupied by electrons of the second and third covalent bonds.

Functional Groups

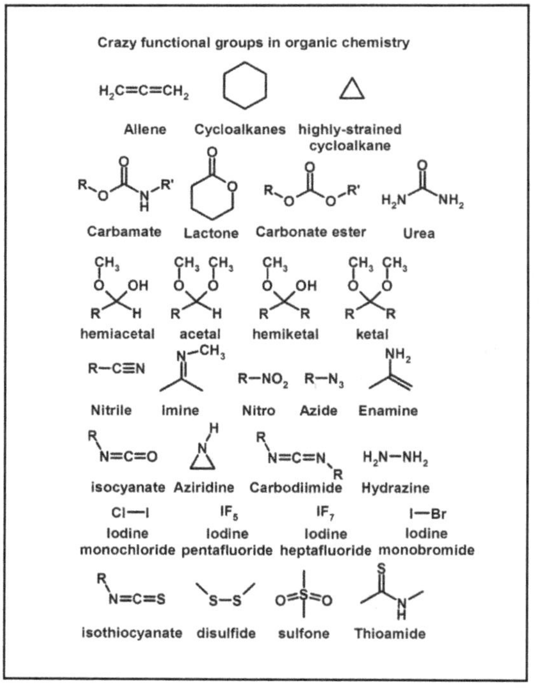

Chemists observed early in the study of organic compounds that certain groups of atoms and associated bonds, known as functional groups, confer specific reactivity patterns on the molecules of which they are a part. Although the properties of each of the several million organic molecules whose structure is known are unique in some way, all molecules that contain the same functional group have a similar pattern of reactivity at the functional group site. Thus, functional groups are a key organizing feature of organic chemistry. By focusing on the functional groups present in a molecule (most molecules have more than one functional group), several of the reactions that the molecule will undergo can be predicted and understood.

Because carbon-to-carbon and carbon-to-hydrogen bonds are extremely strong and the charge of the electrons in these covalent bonds is spread more or less evenly over the bonded atoms, hydrocarbons that contain only single bonds of these two types are not very reactive. The reactivity of a molecule increases if it contains one or more weak bonds or bonds that have an unequal distribution of electrons between the two atoms. If the two electrons of a covalent bond are, for one reason or another, drawn more closely to one of the bonded atoms, that atom will develop a partial negative charge and the atom to which it is bonded will develop a partial positive charge. A covalent bond in which the electron pair linking the atoms is shared unequally is known as a polar bond. Polar bonds, and any other bonds that have unique electronic properties, confer the potential for chemical reaction on the molecule in which they are present. This is because, for every reaction, one or more bonds of a molecule must be broken and new bonds formed. The presence of a partial negative charge (a region of high electron density) will draw to itself other atoms or groups of atoms that are deficient in electron density. This initiates the process of bond breaking that is a prerequisite for a chemical reaction. For these reasons, molecules with regions of increased or decreased electron density are especially important for chemical change.

There are two major bonding features that generate the reactive sites of functional groups. The first, already mentioned, is the presence of multiple bonds. Both double and triple bonds have regions of high electron density lying outside the atom-to-atom bond axis. Double and triple bonds are known as functional groups, a term that is used to identify atoms or groups of atoms within a molecule that are sites of comparatively high reactivity. A second type of reactive site results when an atom other than carbon or hydrogen (termed a *heteroatom*) is bonded to carbon. All heteroatoms have a greater or lesser attraction for electrons than does carbon. Thus, each bond between a carbon and a heteroatom is polar, and the degree of polarity depends on the difference between the electron-attracting properties of the two atoms. The most important atomic groupings that contain such reactive polar bonds are also able to generate functional groups.

To emphasize the generality of reactions between molecules that contain the same functional group, chemists often represent the less reactive portions of a molecule by the symbol R. Thus, all molecules that contain a double bond, however complicated, can be represented by the general formula for an alkene—i.e.,

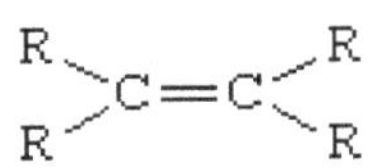

This type of formula suggests that the molecule will undergo those reactions that are common to double bonds and that the reaction will occur at the double bond. The rest of the molecule, represented by the four R groups, will remain unchanged by the reaction occurring at the functional group site.

Molecules with more than one functional group, called poly functional, may have more complicated properties that result from the identity—and interconnectedness—of the multiple functional groups. Many natural products contain several functional groups located at specific sites within a large, complicated, three-dimensional structure.

Alkanes

Alkanes are compounds that consist entirely of atoms of carbon and hydrogen (a class of substances known as hydrocarbons) joined to one another by single bonds. The shared electron pair in each of these single bonds occupies space directly between the two atoms; the bond generated by this shared pair is known as a sigma (σ) bond. Both carbon-carbon and carbon-hydrogen sigma bonds are single strong, nonpolar covalent bonds that are normally the least reactive bonds in organic molecules. Alkane sequences form the inert framework of most organic compounds. For this reason, alkanes are not formally considered a functional group. When a hydrocarbon chain is connected as a substituent to a more fundamental structural unit, it is termed an alkyl group. The simplest examples of alkanes are methane (CH_4; the principal constituent of natural gas), ethane (C_2H_6), propane (C_3H_8; widely used as a barbecue fuel), and butane (C_4H_{10}; the liquid fuel in pocket lighters). Hydrocarbon chains commonly occur in cyclic forms, or rings; the most common example is cyclohexane (C_6H_{12}).

Alkenes

Organic compounds are termed alkenes if they contain a carbon-carbon double bond. The shared electron pair of one of the bonds is a σ bond. The second pair of electrons occupies space on both sides of the σ bond; this shared pair constitutes a pi (π) bond. A π bond forms a region of increased electron density because the electron pair is more distant from the positively charged carbon nuclei than is the electron pair of the σ bond. Even though a carbon-carbon double bond is very strong, a π bond will draw to itself atoms or atomic groupings that are electron-deficient, thereby initiating a process of bond-breaking that can lead to rupture of the π bond and formation of new σ bonds. A simple example of an alkene reaction, which illustrates the way in which the electronic properties of a functional group determine its reactivity, is the addition of molecular hydrogen to form alkanes, which contain only σ bonds.

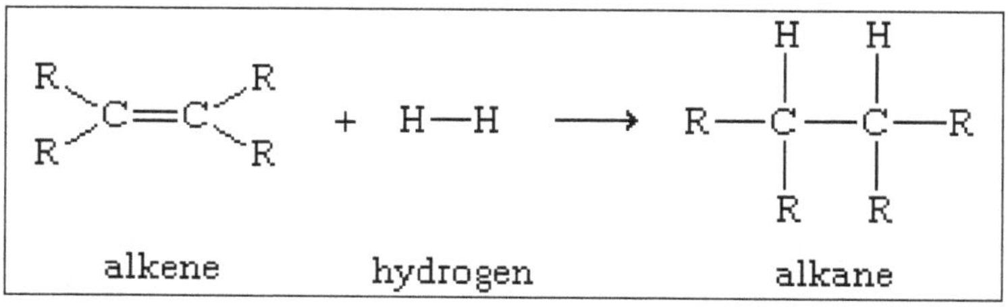

Such reactions, in which the π bond of an alkene reacts to form two new σ bonds, are energetically favourable because the new bonds formed (two carbon-hydrogen σ bonds) are stronger than the bonds broken (one carbon-carbon π bond and one hydrogen-hydrogen σ bond). Because the addition of atoms to the π bond of alkenes to form new σ bonds is a general and characteristic reaction of alkenes, alkenes are said to be unsaturated. Alkanes, which cannot be transformed by addition reactions into molecules with a greater number of σ bonds, are said to be saturated.

The alkene functional group is an important one in chemistry and is widespread in nature. Some common examples (shown here) include ethylene (used to make polyethylene), 2-methyl-1,3-butadieneisoprene (used to make rubber), and vitamin A (essential for vision).

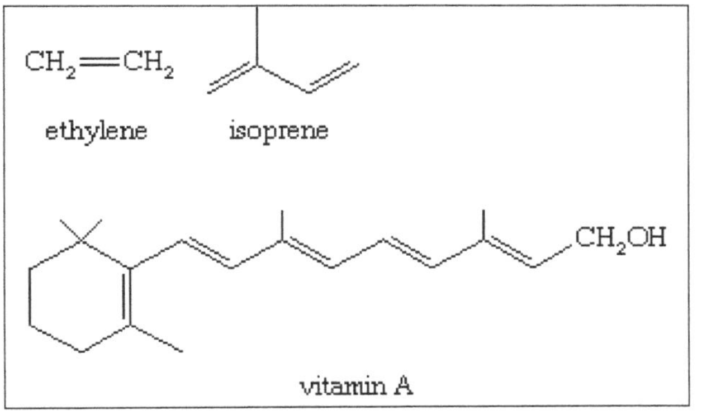

For ethene, both the carbon atoms of an alkene and the four atoms connected to the double bond lie in a single plane.

Alkynes

Molecules that contain a triple bond between two carbon atoms are known as alkynes. The triple bond is made up of one σ bond and two π bonds. As in alkenes, the π bonds constituteregions of increased electron density lying parallel to the carbon-carbon bond axis. Carbon-carbon triple bonds are very strong bonds, but reactions do occur that break the π bonds to form stronger σ bonds.

The most common example of an alkyne is ethyne (also known as acetylene), used as a fuel for oxyacetylene torches in welding applications. Alkynes are not abundant in nature, but the fungicide capillan contains two alkyne functional groups.

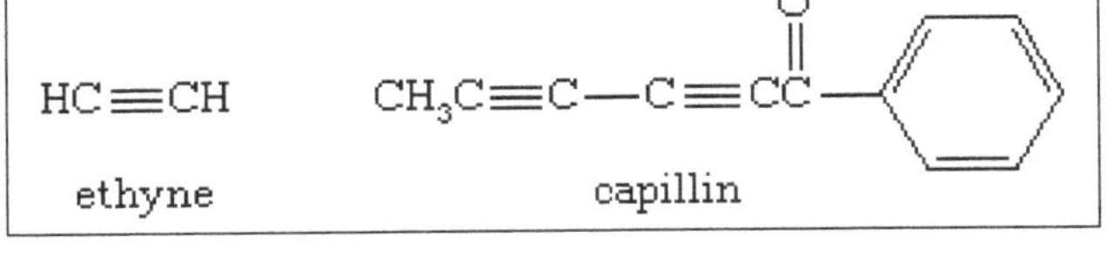

Aromatic Hydrocarbons (Arenes)

A distinctive set of physical and chemical properties is imparted to molecules that contain a functional group composed of three pairs of doubly bonded atoms (usually all carbon atoms) bonded together in the shape of a regular planar (flat) hexagon. The hexagonal ring is usually drawn with an alternating sequence of single and double bonds. The molecule benzene, C_6H_6, first discovered by English physicist and chemist Michael Faraday in 1825, is the smallest molecule that can contain this functional group, and arenes contain one or more benzene (or structurally similar) rings. Because benzene and many larger arenes have a strong odour, they have long been known as aromatic hydrocarbons. Benzene, and all the larger arenes, have a characteristic planar structure forced on them by the electronic requirements of the six (or more) pi electrons. When named as substituents on other structural units, the aromatic units are called aryl substituents. Naphthalene, the active component of mothballs, contains two fused benzene rings. Benzo[a]pyrene, an aromatic hydrocarbon produced in small amounts by the combustion of organic substances, contains five fused benzene rings. Like several other polycyclic aromatic hydrocarbons, it is carcinogenic. Aromatic compounds are widely distributed in nature. Benzaldehyde, anisole, and vanillin, for example, have pleasant aromas.

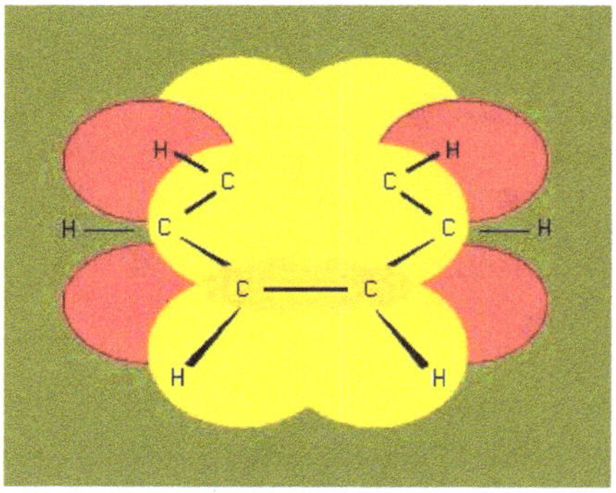

Benzene is the smallest of the organic aromatic hydrocarbons. It contains sigma bonds (represented by lines) and regions of high-pi electron density, formed by the overlapping of *p* orbitals (represented by the dark yellow shaded area) of adjacent carbon atoms, which give benzene its characteristic planar structure.

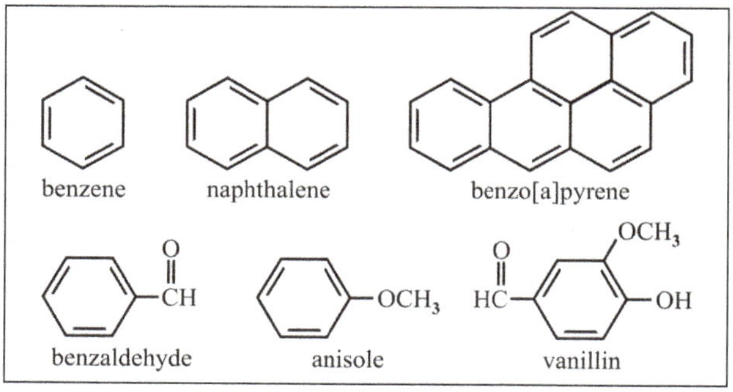

Alcohols and Phenols

An oxygen atom normally forms two σ bonds with other atoms; the water molecule, H_2O, is the simplest and most common example. If one hydrogen atom is removed from a water molecule, a hydroxyl functional group (−OH) is generated. When a hydroxyl group is joined to an alkane framework, an alcohol such as ethanol, is produced.

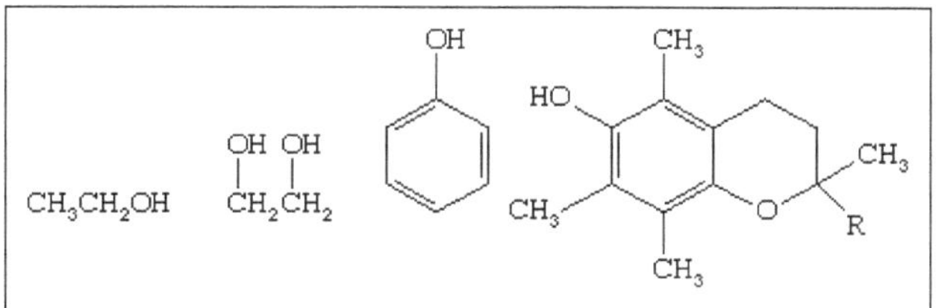

ethanol ethylene glycol phenol vitamin E

When the hydroxyl group is joined to an aryl ring, a phenol results (shown above). Both alcohols and phenols are widespread in nature, with alcohols being especially ubiquitous. The hydroxyl group of alcohols and phenols is responsible for an interesting variety of physical and chemical properties. The biochemical action of vitamin E, for example, depends largely on the reactivity of the phenol functional group.

An oxygen atom is much more electronegative than carbon or hydrogen atoms, so both carbon-oxygen and hydrogen-oxygen bonds are polar. The oxygen atom is slightly negatively charged, and the carbon and hydrogen atoms are slightly positively charged. The polar bonds of the hydroxyl group are responsible for the major reaction characteristics of alcohols and phenols. In general, these reactions are initiated by reaction of electron-deficient groups with the negatively charged oxygen atom or by reaction of electron-rich groups with the positively charged atoms—namely, carbon or hydrogen—bonded to oxygen.

Ethers and Epoxides

An organic molecule in which an oxygen atom is bonded to two carbon atoms through two sigma bonds is known as an ether. Ether molecules occur widely in nature. Diethyl ether was once widely used as an anesthetic. An aromatic ether known as Nerolin II (2-ethoxynaphthalene) is used in perfumes to impart the scent of orange blossoms. Cyclic ethers, such as tetrahydrofuran, are commonly used as organic solvents. Although ethers contain two polar carbon-oxygen bonds, they are much less reactive than alcohols or phenols.

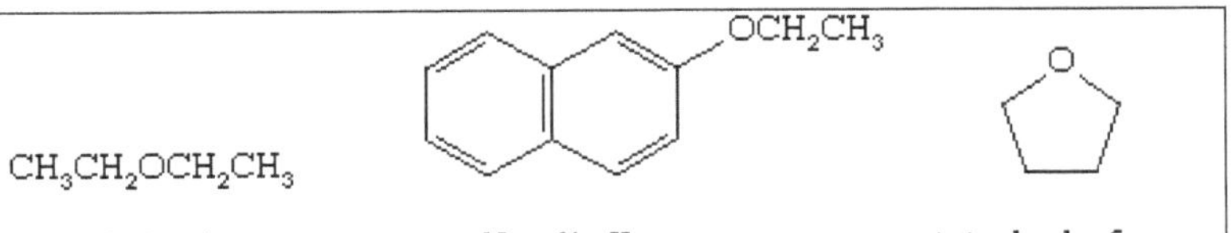

diethyl ether Nerolin II tetrahydrofuran

Epoxides are cyclic ethers that contain a three-membered ring. The simplest example is oxirane (ethylene oxide). An epoxide is one of the functional groups in the insect hormone known as juvenile hormone.

Thiols

A thiol is structurally similar to an alcohol but contains a sulfur atom in place of the oxygenatom normally found in an alcohol. The outstanding feature of thiols is their foul smell. The simplest thiol is hydrogen sulfide, H_2S, the sulfur analog of water. It can be detected by the human nose at a concentration of a few parts per billion and is readily identifiable as having a rotten-egg odour. Ethanethiol is added in trace amounts to natural gas to give it a detectable odour, and striped skunks deter predators by releasing a liquid spray containing 3-methyl-1-butanethiol. When present as a substituent on another structural unit, the SH group is commonly termed *mercapto*, as in 2-mercaptoethanol.

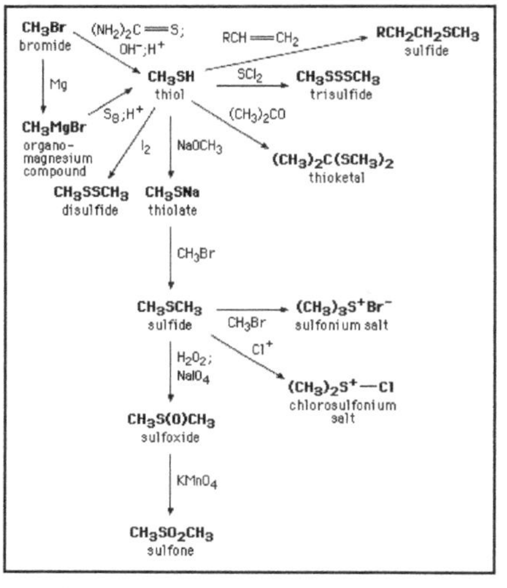

Thiols are organic functional groups, similar in structure to alcohols but containing sulfur atoms in place of oxygen atoms. Thiols react to form sulfide compounds and other sulfur-containing compounds.

Amines

Amines are functional group compounds that contain at least one nitrogen atom bonded to hydrogen atoms or to alkyl or aryl groups. If the substituents (other than hydrogen atoms) are alkyl groups, the resulting compounds are termed *alkyl amines*. If one or more substituents is an aryl group, the compounds are termed *aryl amines*. Amines are commonly categorized as primary, secondary, or tertiary, depending on whether the nitrogen atom is bonded to one, two, or three alkyl or aryl groups, respectively. The nitrogen atom is bonded to its hydrogen atoms and alkyl groups by sigma (σ) bonds, but the nitrogen atom also bears a nonbonded electron pair. The three σ bonds and nonbonded electron pair are oriented around the nitrogen atom in a distorted tetrahedral geometry.

In some compounds, the non-bonded electron pair on the nitrogen atom is replaced by a fourth σ bond to a hydrogen atom or to an alkyl or aryl group. The resulting compound, called a quaternary ammonium salt, has a positive charge on the nitrogen atom and a tetrahedral arrangement of groups around the nitrogen atom. Amines are very common organic molecules, and many are physiologically active. Amphetamine, for example, is a central nervous system stimulant and acts as an antidepressant. Amines are particularly valuable because of their ability to act as bases, a property that is a consequence of the ability of amines to accept hydrogen atoms from acidic molecules.

Halides

Halides, or organohalides, are compounds that contain a halogen atom (fluorine, chlorine, bromine, or iodine) bonded to a carbon atom by a polar bond. The slightly positive charge that exists on the carbon atom in carbon-halogen bonds is the source of reactivity of halides. A wide variety of organohalides have been discovered in marine organisms, and several simple halide compounds have important commercial applications. Chloroethane (ethyl chloride) is a volatile liquid that is used as a topical anesthetic. Chloroethene (vinyl chloride) is the monomeric building block for polyvinyl chloride (PVC), and the mixed organohalide halothane is an inhalation anesthetic. The compound epibatidine, isolated from glands on the back of an Ecuadorian poison frog, has been found to be an especially potent painkiller.

Aldehydes and Ketones

When an oxygen atom forms a double bond to a carbon atom, a carbonyl functional group is obtained. The carbon atom of a carbonyl group is bonded to two other atoms in addition to the oxygen atom. A wide range of functional groups are produced by the presence of different atomic groupings on the carbon of the carbonyl group. Two of the most important are aldehydes and ketones. In a ketone, both atoms bonded to the carbonyl carbon are other carbon atoms, and, in an aldehyde, at least one atom on the carbonyl carbon is a hydrogen. Similar to the double bond of alkenes, the carbon-oxygen double bond is made up of a σ bond, whose electron pair lies between the bonded atoms, and a π bond, whose electron pair occupies space on both sides of the σ bond.

Many aldehydes and ketones have pleasant, fruity aromas, and these compounds are frequently responsible for the flavour and smell of fruits and vegetables. A 40 percent solution of formaldehyde in water is formalin, a liquid used for preserving biological specimens. Benzaldehyde is an aromatic aldehyde and imparts much of the aroma to cherries and almonds. Butanedione, a ketone with two carbonyl groups, is partially responsible for the odour of cheeses. Civetone, a large cyclic ketone, is secreted by the civet cat and is a key component of many expensive perfumes.

The carbonyl group has a wide variety of reaction pathways open to it. Because of its π bond, the carbonyl group undergoes addition reactions similar to those that occur with alkenes but with a few important differences. Whereas carbon-carbon double bonds are nonpolar, carbon-oxygen double bonds are polar. Species that add to a carbonyl group to form new σ bonds react in such a way that electrophilic (electron-seeking) groups attack the oxygen atom and nucleophilic groups (those seeking positively charged centres) attack the carbon atom. Furthermore, addition to a carbonyl group results in the breaking of a strong π bond. The energy relationships of carbonyl addition reactions are consequently very different from those of alkene addition reactions. Other reaction possibilities of carbonyl compounds depend on the nature of the atomic groupings, termed *substituents*, attached to the carbonyl carbon. When both substituents are unreactive alkane fragments, as in ketones, there are few reactions other than carbonyl additions. When one of the substituents is not an alkane fragment, different possibilities emerge. In aldehydes, the carbonyl carbon is bonded to a hydrogen atom, and reactions that involve this hydrogen atom distinguish the reactions of aldehydes from those of ketones.

Carboxylic Acids

The conjunction of a carbonyl and a hydroxyl group forms a functional group known as a carboxyl group.

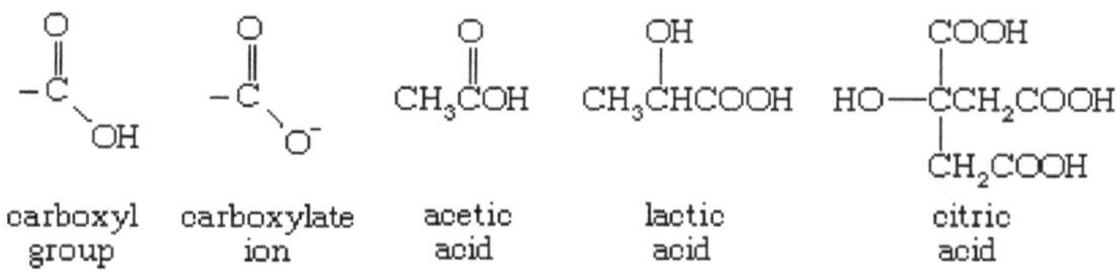

The hydrogen of a carboxyl group can be removed (to form a negatively charged carboxylate ion), and thus molecules containing the carboxyl group have acidic properties and are generally known as carboxylic acids. Vinegar is a 5 percent solution of acetic acid in water, and its sharp acidic taste is due to the carboxylic acid present. Lactic acid provides much of the sour taste of pickles and sauerkraut and is produced by contracting muscles. Citric acid is a major flavour component of citrus fruits, such as lemons, grapefruits, and oranges. Ibuprofen, an effective analgesic and anti-inflammatory agent, contains a carboxyl group.

The structural unit containing an alkyl group bonded to a carbonyl group is known as an acyl group. A family of functional groups, known as carboxylic acid derivatives, contains the acyl group bonded to different substituents.

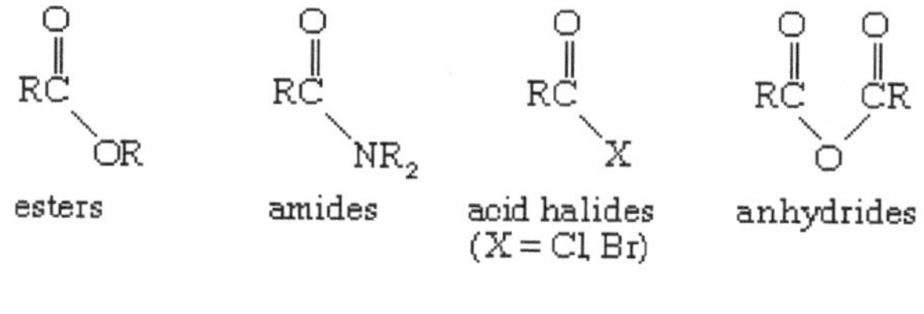

Esters have an alkoxy (OR) fragment attached to the acyl group; amides have attached amino groups ($-NR_2$); acyl halides have an attached chlorine or bromine atom; and anhydrides have an attached carboxyl group. Each type of acid derivative has a set of characteristic reactions that qualifies it as a unique functional group, but all acid derivatives can be readily converted to a carboxylic acid under appropriate reaction conditions. Many simple esters are responsible for the pleasant odours of fruits and flowers. Methyl butanoate, for example, is present in pineapples. Urea, the major organic constituent of urine and a widely used fertilizer, is a double amide of carbonic acid. Acyl chlorides and anhydrides are the most reactive carboxylic acid derivatives and are useful chemical reagents, although they are not important functional groups in natural substances.

Polyfunctional Compounds

Although each of the functional groups introduced above has a characteristic set of favoured reactions, it is not always possible to predict the properties of organic compounds that contain several different functional groups. In polyfunctional organic compounds, the functional groups often interact with one another to impart unique reactivity patterns to the compounds. As chemistry evolves as a science, it becomes possible to understand more of the behaviour of complex molecules, and chemists are able to design laboratory syntheses of increasingly complicated molecules, basing the synthetic plan upon the reactivity trends of functional groups.

States of Matter

The matter is made up of very tiny particles and these particles are so small that we cannot see them with naked eyes. It has been observed that matter exists in nature in different forms. Some substances are rigid and have a fixed shape like wood and stone; some substances can flow and take the shape of their container like water, while there are forms of matter that do not have definite shape or size such as air. Thus, the matter can be classified into different categories based on the physical properties exhibited by them and the states in which they exist; these are called states of matter.

Following are the basic three states of matter:

- Solid
- Liquid
- Gas

Apart from the three, there are 2 more states of matter which we do not see in our everyday life. They are Plasma & Bose-einstein condensate.

Solid

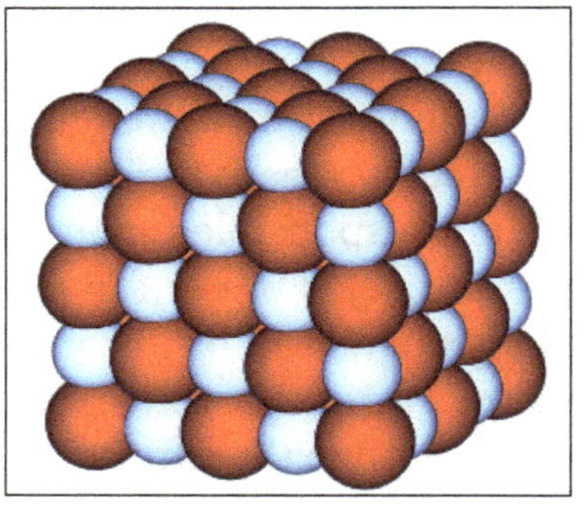

Structure of solids

- In solids, particles are tightly or closely packed.
- The gaps between the particles are tiny and hence it is tough to compress them.
- Solid has a fixed shape and volume.
- Due to its rigid nature, particles in solid can only vibrate about their mean position and cannot move.
- Force of attraction between particles is adamant.
- The rate of diffusion in solids is very low.
- An example of solids: solid ice, sugar, rock, wood, etc.

Liquid

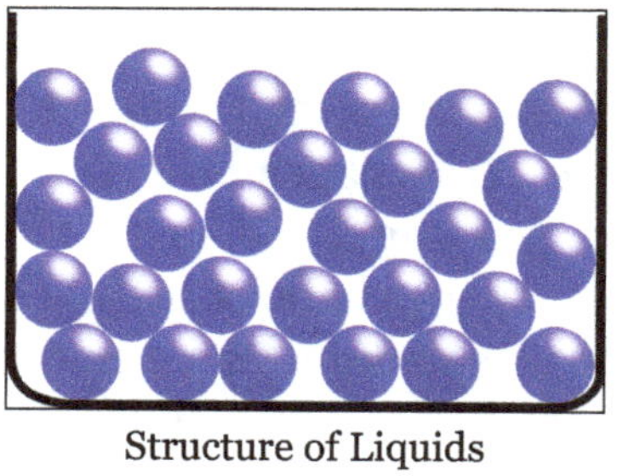

Structure of Liquids

- In a liquid state of matter, particles are less tightly packed as compared to solids.
- Liquids take the shape of the container in which they are kept.
- Liquids are difficult to compress as particles have less space between them to move.
- Liquids have fixed volume but no fixed shape.
- The rate of diffusion in liquids is higher than that of solids.

- Force of attraction between the particles is weak than solids.
- Example of a liquid state of matter: water, milk, blood, coffee, etc.

Gas

- In gases, particles are far apart from each other.
- Force of attraction between the particles is negligible and they can move freely.
- Gases have neither a fixed volume nor a fixed shape.
- The gaseous state has the highest compressibility as compared to solids and liquids.
- The rate is diffusion is higher than solids and liquids.
- The kinetic energy of particles is higher than in solids and liquids.
- An example of gases: air, helium, nitrogen, oxygen, carbon dioxide, etc.

Chemical Equations and Stoichiometry

Chemical Equation

A chemical equation is something you will encounter every day in chemistry. It's a written representation, using numbers and symbols, of the process that occurs during a chemical reaction.

Way to Write a Chemical Equation

A chemical equation is written with the reactants on the left side of an arrow and the products of the chemical reaction on the right. The head of the arrow typically points toward the right or the product side of the equation, although some equations may indicate equilibrium with the reaction proceeding in both directions simultaneously.

The elements in an equation are denoted using their symbols. Coefficients next to the symbols indicate the stoichiometric numbers. Subscripts are used to indicate the number of atoms of an element present in a chemical species.

An example of a chemical equation may be seen in the combustion of methane:

$$CH_4 + 2O_2 \rightarrow CO_2 + 2H_2O$$

Participants in the Chemical Reaction: Element Symbols

The symbols for the elements to understand what is taking place in a chemical reaction. In this reaction, C is carbon, H is hydrogen, and O is oxygen.

Left Side of the Equation: Reactants

The reactants in this chemical reaction are methane and oxygen: CH_4 and O_2.

Right Side of the Equation: Products

The products of this reaction are carbon dioxide and water: CO_2 and H_2O.

Direction of Reaction: Arrow

It is the convention to place the reactants on the left side of a chemical equation and the products on the right side. The arrow between the reactants and products should point from left to right or if the reaction is proceeding both ways, point in both directions (this is common). If your arrow points from right to left, it's a good idea to re-write the equation the conventional way.

Balancing Mass and Charge

Chemical equations may be either unbalanced or balanced. An unbalanced equation lists the reactants and products, but not the ratio between them. A balanced chemical equation has the same number and types of atoms on both sides of the arrow. If ions are present, the sum of the positive and negative charges on both sides of the arrow is also the same.

Indicating States of Matter

It's common to indicate the state of matter in a chemical equation by including parentheses and an abbreviation right after a chemical formula. This can be seen in the following equation:

$$2H_2(g)+O_2(g)\rightarrow 2H_2O$$

Hydrogen and oxygen are indicated by (g), which means they are gases. Water is marked (in above equation), which means it is a liquid. Another symbol you may see is (aq), which means the chemical species is in water—or an aqueous solution. The (aq) symbol is a sort of shorthand notation for aqueous solutions so that water doesn't have to be included in the equation. It's particularly common when ions are present in a solution.

Stoichiometry

Stoichiometry is the study of the quantitative relationships or ratios between two or more substances undergoing a physical change or chemical change (chemical reaction).

Most often, stoichiometry calculations deal with the mass or volumes of products and reactants.

Stoichiometric Coefficient

Stoichiometric coefficient (v) is the number appearing before the symbol for each compound in the equation for a chemical reaction. By convention, it is negative for reactants and positive for products.

Stoichiometric coefficients describe the stoichiometry of the chemical reaction.

$$aA + bB \rightleftarrows cC + dD$$

In this equation, a, b, c and d are called as Stoichiometric coefficients of the A, B, C and D respectively.

Isotopes, Isobars and Isotones

Isotopes

- These are elements which have the same atomic number but different atomic mass.
- They have the same atomic number because the number of protons that are inside their nuclei remains the same.
- But, they have different atomic mass because the number of neutrons that are also inside their nuclei is different.
- As the number of protons inside nuclei remains same, therefore the overall charge of the elements also remains same as in isotopes: no of protons = no of electrons.
- Hence, as isotopes overall charge remains neutral, therefore their chemical properties will also remain identical.

Therefore, Isotopes are chemically same but physically different.

Examples of isotopes are:

1. Isotopes of hydrogen

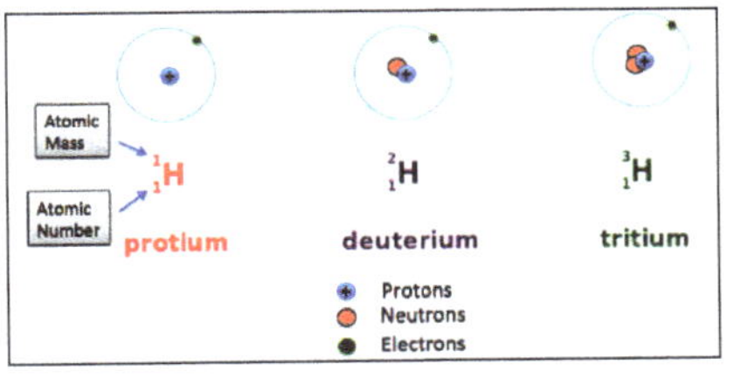

2. Isotopes of carbon

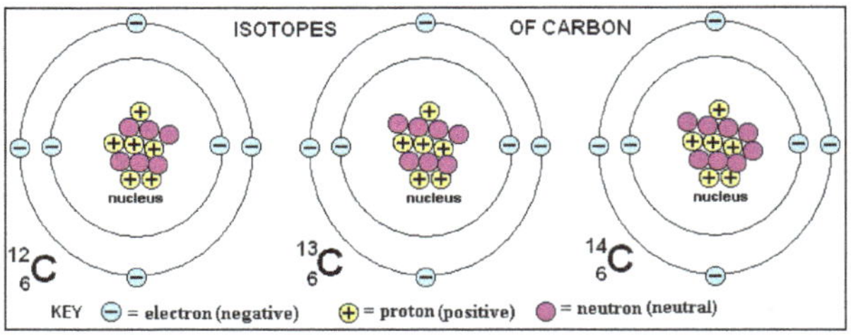

Isobars

- These are elements which have same atomic mass but a different atomic number.
- Hence, as isobars have different atomic number, therefore, they have a different number of electrons, and their chemical properties are different.
- The light nuclei have unstable isobars.
- Heavy nuclei have stable isobars and these occur in pairs.
- If the number of protons of one isobar matches with that of another, then they are called as mirror-nuclides of each other.

Therefore, Isobars are chemically different but physically same.

examples:

$$C_a = C_t V_t M / V_a$$

$$_{26}Fe^{58}, {}_{27}Ni^{58}$$

Isotones

These are elements having the same number of neutrons.

Example:

Oxygen $^{16}_{8}O(p=8;n=8)$

Nitrogen $^{15}_{7}N(p=7;n=8)$

Carbon $^{14}_{6}C(p=6;n=8)$

are isotones because of having same no. of neutron 8.

Gas Laws

All gases generally show similar behaviour when the conditions are normal. But with a slight change in physical conditions like pressure, temperature or volume these show a deviation. Gas laws are an analysis of this behaviour of gases. The variables of state like the Pressure, Volume and Temperature of a gas depict its true nature. hence gas laws are relations between these variables.

Boyle's Law

Boyle's law states the relation between volume and pressure at constant temperature and mass. Robert Boyle conducted an experiment on gases to study the deviation of its behaviour in changed physical conditions.

It states that under a constant temperature when the pressure on a gas increases its volume decreases. In other words according to Boyle's law volume is inversely proportional to pressure when the temperature and the number of molecules are constant.

$$p \alpha 1/V$$
$$p = k_1 1/V$$

k_1 here is a proportionality constant,

V is the Volume and p is the pressure.

On rearranging, we get: $k_1 = pV$

Now, if a fixed mass of gas undergoes an expansion at constant temperature then the final volume and pressure shall be P_2 and V_2. The initial volume and initial pressure here is P_1 and V_1 then according to Boyle's law: $P_1 \times V_1 = p_2 \times V_2 = cons\tan t(k1)$

$$P_1/P_2 = V_2/V_1$$

So according to Boyle's law, if the pressure is doubled then at constant temperature the volume of that gas is reduced to half. The reason being the intermolecular force between the molecules of the gaseous substance. In a free state, a gaseous substance occupies a larger volume of the container due to the scattered molecules.

When a pressure is applied to the gaseous substance, these molecules come closer and occupy a lesser volume. In other words, the pressure applied is directly proportional to the density of the gas. Boyle's law can be graphically represented as follows:

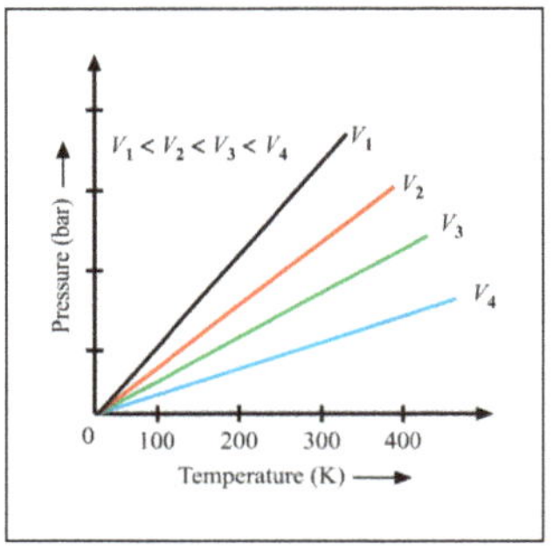

Charle's Law

Jacques Charles in 1787 analyzed the effect of temperature on the volume of a gaseous substance at a constant pressure. He did this analysis to understand the technology behind the hot air balloon flight. According to his findings, at constant pressure and for constant mass, the volume of a gas is directly proportional to the temperature.

This means that with the increase in temperature the volume shall increase while with decreasing temperature the volume decreases. In his experiment, he calculated that the increase in volume with every degree equals 1/273.15 times of the original volume. Therefore, if the volume is V_0 at 0° C and V_t is the volume at t° C then,

$$V_t = V_0 + t/273.15V_0 \Rightarrow V_t = V_0(1 + t/273.15)$$

$$\Rightarrow V_t = V_0(273.15 + t/273.15)$$

For the purpose of measuring the observations of gaseous substance at temperature 273.15 K, we use a special scale called the Kelvin Temperature Scale. The observations of temperature (T) on this scale is 273.15 greater than the temperature (t) of the normal scale.

$$T = 273.15 + t$$

While, when T = 0° c then the reading on the Celsius scale is 273.15. The Kelvin Scale is also called Absolute Temperature Scale or Thermodynamic Scale. This scale is used in all scientific experiments and works. In the equation $\left[V_t = V_o\left(273.15 + t / 273.15\right)\right]$ if we take the values T_t = 273.15+t and T_o = 273.15 then:

$$V_t = V_o\left(T_t / T_o\right)$$

Which implies $V_t / V_o = \left(T_t / T_o\right)$, which can also be written as,

$$V_2 / V_1 = T_2 / T_1$$
$$or\ V_1 / T_1 = V_2 / T_2$$
$$V / T = constant = k_2$$
$$Therefore, V = k_2 T$$

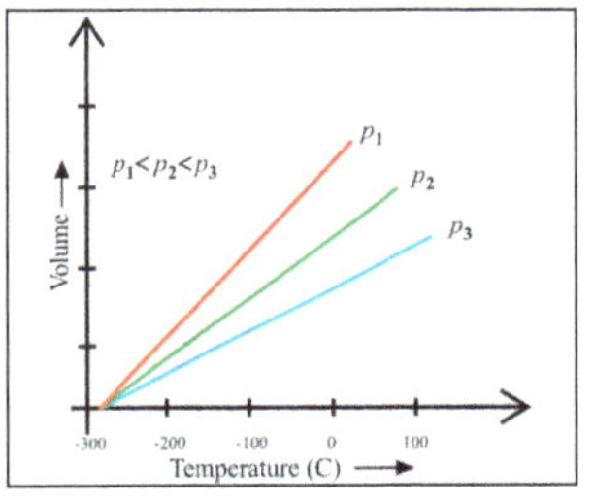

The graphical representation of Charles law is shown in the figure above. Its an isobar graph as the pressure is constant with volume and temperature changes under observation.

Gay-Lussac's law

Also referred to as Pressure-Temperature Law, Gay Lussac's Law was discovered in 1802 by a French scientist Joseph Louis Gay Lussac. While building an air thermometer, Gay-Lussac accidentally discovered that at fixed volume and mass of a gas, the pressure of that gas is directly proportional to the temperature. This mathematically can be written as: $p \alpha T$

$$\Rightarrow p / T = \text{constant} = k_3$$

The temperature here is measured on the Kelvin scale. The graph for the Gay Lussac's Law is called as an isochore because the volume here is constant.

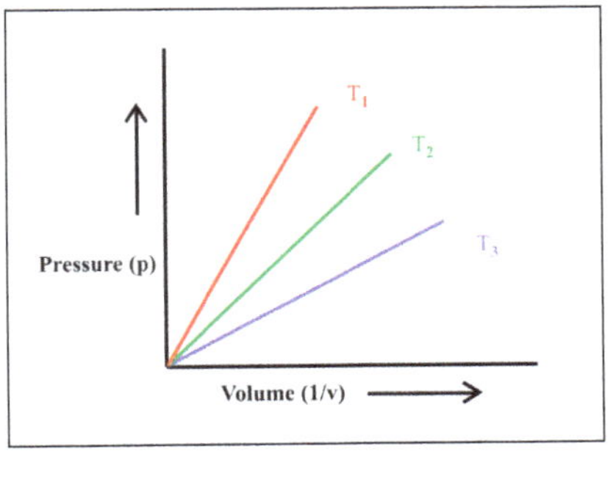

Avogadro's Law

Amedeo Avogadro in 1811 combined the conclusions of Dalton's Atomic Theory and Gay Lussac's Law to give another important Gas law called the Avogadro's Law. According to Avogadro's law, at constant temperature and pressure, the volume of all gases constitutes an equal number of molecules. In other words, this implies that in unchanged conditions of temperature and pressure the volume of any gas is directly proportional to the number of molecules of that gas.

Mathematically, $V \alpha n$

Here, n is the number of moles of the gas.

Hence, $V = k_4 n$

The number of molecules in a mole of any gas is known as the Avogadro's constant and is calculated to be 6.022×10^{23}. The values for temperature and pressure here are the standard values. For temperature, we take it to be 273.15 K while for the pressure it equals 1 bar or 10^5 pascals. At these Standard Temperature Pressure (STP) values, one mole of a gas is supposed to have the same volume.

Now, $n = m / M$

According to Avogadro's equation: $V = k_4 (m / M)$

$$M = k_4 (m / V)$$

$$m / V = d(density); Therefore\, M = k_4 D$$

This means that at an unchanged temperature and pressure conditions, the molar mass of every gas is directly proportional to its density.

Titration

Titration is the process in which one solution is added to another solution such that it reacts under conditions in which the added volume may be accurately measured. It is used in quantitative analytical chemistry to determine an unknown concentration of an identified analyte. Titrations are most commonly associated with acid-base reactions, but they may involve other types of reactions as well.

Titration is also known as titrimetry or volumetric analysis. The chemical of unknown concentration is called the analyte or titrand. A standard solution of a reagent of known concentration is called the titrant or titrator. The volume of titrant that is reacted (usually to produce a color change) is called the titration volume.

Working of Titration

A typical titration is set up with an Erlenmeyer flask or beaker containing a precisely known volume of analyte (unknown concentration) and a color-change indicator. A pipette or burette containing a known concentration of titrant is placed above the flask or beaker of an analyte. The starting volume of the pipette or burette is recorded. Titrant is dripped into the analyte and indicator solution until the reaction between titrant and analyte is complete, causing a color change (the endpoint). The final volume of the burette is recorded, so the total volume used can be determined.

The concentration of analyte may then be calculated using the formula,

$$C_a = C_t V_t M / V_a$$

Where,

- C_a is the analyte concentration, usually in molarity;
- C_t is the titrant concentration, in the same units;
- V_t is the volume of titrant, usually in liters;
- M is the mole ratio between the analyte and reactant from the balanced chemical equation;
- V_a is the volume of an analyte, usually in liters.

Titration Curve

Titration curves are graphs that display the information gathered by a titration. The information is displayed on a two-dimensional axis, typically with chemical volume on the horizontal axis and solution pH on the vertical axis. The curve of the graph shows the change in solution pH as the volume of the chemical changes due to the addition of the titrant. Chemical volume is an independent variable on the graph, while pH (which changes as the volume of the measured chemical increases) is a dependent variable.

Reading a Titration Graph

Titration curves show how the pH of the solution changes as a known chemical is added to the solution, so any point along the curve gives you information on solution pH as the volume of the known chemical increases. The graph typically shows a gradual increase of pH until nearly all of the hydrogen ions in the solution are neutralized; at this point, the graph increases sharply. This almost vertical movement continues until the solution begins turning basic, at which point the graph levels out again. Almost all titration curves of acids or bases follow this same shape.

Using the graph, the point at which neutralization occurs (indicated by the steep shift of the curve) can be found, as can the equivalence point where the solution reaches an equilibrium (which is halfway up the steep slope). The point at which the pH levels off again is also clearly visible. Using these points (and occasionally other select points along the line) and other data about the chemicals involved, it's possible to calculate the concentration of the known chemical within the analyte.

Types of Titrations

During titration experiments, a scientist will have a reagent--a solution with a known chemical make-up and concentration--and a sample. The scientist may have a general idea of what the sample is, but needs to know the concentration of a chemical in the sample. For example, when a scientist wants to find out how much of a certain pollutant is in drinking water. In a basic titration experiment, the reagent is added to the sample until a specific reaction occurs. This reaction could be a color change, a pH change or the precipitation of a chemical in the sample--precipitation is when a reaction occurs and causes a solid to form within a liquid. The amount of reagent used to cause the reaction, tells the scientist the concentration of the sought-after chemical in the sample.

Acid-Base Titration

To determine the content of a specific acid such as hydrochloric acid or a base, such as sodium hydroxide, in a liquid, chemists opt for acid-base titration. When analyzing a solution for acid, the process is called acidimetry; when analyzing for a base it is called alkalimetry. In this type of titration a reagent is added until the sample solution reaches a specified pH level. This type of titration relies on a pH meter or a dye to track the change in pH. Like litmus paper, the dye will change to a certain color once the correct pH has been reached.

Oxidation-reduction Titration

Also known a redox titration, this form of titration relies on a gain or loss of electrons within a sample to find out what is in the sample. Redox titration can be used to study contamination in drinking water or the concentration of metals within a solution. This type of titration has many names depending on the substance used to cause the observable change during titration. For example, in permanganate titrations, potassium permanganate - a form of salt - causes a reaction that can show how much hydrogen peroxide is in a sample.

Precipitation Titration

In precipitation titration, a reagent is added to a sample until a reaction occurs that causes a solid to precipitate from -- or "fall out" -- of the sample. Precipitation titration can determine the amount of salts in a solution, how much chloride is in drinking water and the amount of specific metals within a sample. This is another form of titration that can have different names depending on the reagent being used. For example, argentometric titrations use silver nitrate -- silver's Latin name is "argentum." When the silver nitrate is added to a sample that contains sodium chloride, a reaction occurs causing white solids of silver chloride to precipitate from the solution.

Complexometric Titrations

This type of titration is similar to precipitation titration in that a solid precipitates out of the sample when a reagent is added. The difference is that in complexometric titration, the solid is formed more quickly and more completely than in precipitation titration, which reduces errors in measurement. Ethylenediaminetetraacetic acid, an acidic powder better known as EDTA, is commonly used in this type of titration because it readily bonds with metals. This type of titration can be used to measure the ingredients within soaps and detergents.

References

- Chemistry, science: britannica.com, Retrieved 3 March, 2019
- Elements-and-compounds: lumenlearning.com, Retrieved 8 June, 2019
- Classification-of-compounds, chemical-compound, science: britannica.com, Retrieved 15 January, 2019
- Matter-solid-liquid-gas, chemistry: byjus.com, Retrieved 12 August, 2019
- What-is-a-chemical-equation: thoughtco.com, Retrieved 23 April, 2019
- Definition-of-stoichiometry: thoughtco.com, Retrieved 28 January, 2019
- Isotopes-isobars-and-isotones: highschoolpedia.com, Retrieved 6 May, 2019
- Gas-laws, states-of-matter, chemistry: toppr.com, Retrieved 18 July, 2019
- Titration-definition: thoughtco.com, Retrieved 29 May, 2019
- What-is-the-titration-curve: sciencing.com, Retrieved 19 March, 2019
- Types-titration: sciencing.com, Retrieved 20 February, 2019

2

Atoms and Molecules

The smallest unit of matter with the properties of chemical element is known as an atom. It comprises of a nucleus and one or more than one electron bound to the nucleus. Molecule is a group of atoms that is bounded together by the chemical bonds. This chapter has been carefully written to provide an easy understanding of the varied facets of atoms and molecules.

Atoms

Atom is smallest unit into which matter can be divided without the release of electrically charged particles. It also is the smallest unit of matter that has the characteristic properties of a chemical element. As such, the atom is the basic building block of chemistry.

Most of the atom is empty space. The rest consists of a positively charged nucleus of protons and neutrons surrounded by a cloud of negatively charged electrons. The nucleus is small and dense compared with the electrons, which are the lightest charged particles in nature. Electrons are attracted to any positive charge by their electric force; in an atom, electric forces bind the electrons to the nucleus.

Because of the nature of quantum mechanics, no single image has been entirely satisfactory at visualizing the atom's various characteristics, which thus forces physicists to use complementary pictures of the atom to explain different properties. In some respects, the electrons in an atom behave like particles orbiting the nucleus. In others, the electrons behave like waves frozen in position around the nucleus. Such wave patterns, called orbitals, describe the distribution of individual electrons. The behaviour of an atom is strongly influenced by these orbital properties, and its chemical properties are determined by orbital groupings known as shells.

The fundamental properties of the atom and its constituent particles and forces are described particles.

Atomic Model

Most matter consists of an agglomeration of molecules, which can be separated relatively easily. Molecules, in turn, are composed of atoms joined by chemical bonds that are more difficult to break. Each individual atom consists of smaller particles—namely, electrons and nuclei. These particles are electrically charged, and the electric forces on the charge are responsible for holding the atom together. Attempts to separate these smaller constituent particles require ever-increasing amounts of energy and result in the creation of new subatomic particles, many of which are charged.

An atom consists largely of empty space. The nucleus is the positively charged centre of an atom and contains most of its mass. It is composed of protons, which have a positive charge, and neutrons, which have no charge. Protons, neutrons, and the electrons surrounding them are long-lived particles present in all ordinary, naturally occurring atoms. Other subatomic particles may be found in association with these three types of particles. They can be created only with the addition of enormous amounts of energy, however, and are very short-lived.

All atoms are roughly the same size, whether they have 3 or 90 electrons. Approximately 50 million atoms of solid matter lined up in a row would measure 1 cm (0.4 inch). A convenient unit of length for measuring atomic sizes is the angstrom (Å), defined as 10−10 metre. The radius of an atom measures 1–2 Å. Compared with the overall size of the atom, the nucleus is even more minute. It is in the same proportion to the atom as a marble is to a football field. In volume the nucleus takes up only 10−14 metres of the space in the atom—i.e., 1 part in 100,000. A convenient unit of length for measuring nuclear sizes is the femtometre (fm), which equals 10−15 metre. The diameter of a nucleus depends on the number of particles it contains and ranges from about 4 fm for a light nucleus such as carbon to 15 fm for a heavy nucleus such as lead. In spite of the small size of the nucleus, virtually all the mass of the atom is concentrated there. The protons are massive, positively charged particles, whereas the neutrons have no charge and are slightly more massive than the protons. The fact that nuclei can have anywhere from 1 to nearly 300 protons and neutrons accounts for their wide variation in mass. The lightest nucleus, that of hydrogen, is 1,836 times more massive than an electron, while heavy nuclei are nearly 500,000 times more massive.

Basic Properties

Atomic Number

The single most important characteristic of an atom is its atomic number (usually denoted by the letter Z), which is defined as the number of units of positive charge (protons) in the nucleus. For example, if an atom has a Z of 6, it is carbon, while a Z of 92 corresponds to uranium. A neutral atom has an equal number of protons and electrons so that the positive and negative charges exactly balance. Since it is the electrons that determine how one atom interacts with another, in the end it is the number of protons in the nucleus that determines the chemical properties of an atom.

Atomic Mass and Isotopes

The number of neutrons in a nucleus affects the mass of the atom but not its chemical properties. Thus, a nucleus with six protons and six neutrons will have the same chemical properties as a nucleus with six protons and eight neutrons, although the two masses will be different. Nuclei with the same number of protons but different numbers of neutrons are said to be isotopes of each other. All chemical elements have many isotopes.

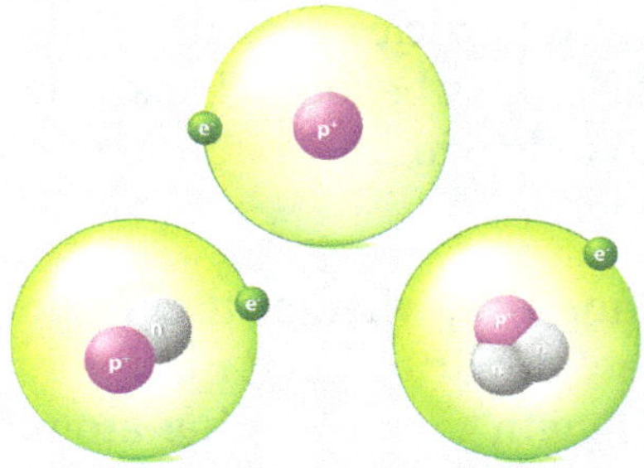

Isotopes of hydrogen: The image shows the three isotopes of the element hydrogen.

All three forms have one proton (pink) and one electron (dark green) but differ in the number of neutrons (gray) in the nucleus. Protium, or ordinary hydrogen (top), has no neutrons. Deuterium, or hydrogen-2 (bottom left), has one neutron. Tritium, or hydrogen-3 (bottom right), has two neutrons.

It is usual to characterize different isotopes by giving the sum of the number of protons and neutrons in the nucleus—a quantity called the atomic mass number. In the above example, the first atom would be called carbon-12 or ^{12}C (because it has six protons and six neutrons), while the second would be carbon-14 or ^{14}C.

The mass of atoms is measured in terms of the atomic mass unit, which is defined to be $^1/_{12}$ of the mass of an atom of carbon-12, or $1.660538921 \times 10^{-24}$ gram. The mass of an atom consists of the mass of the nucleus plus that of the electrons, so the atomic mass unit is not exactly the same as the mass of the proton or neutron.

The Electron

Charge, Mass and Spin

Scientists have known since the late 19th century that the electron has a negative electric charge. The value of this charge was first measured by the American physicist Robert Millikan between 1909 and 1910. In Millikan's oil-drop experiment, he suspended tiny oil drops in a chamber containing an oil mist. By measuring the rate of fall of the oil drops, he was able to determine their weight. Oil drops that had an electric charge (acquired, for example, by friction when moving through the air) could then be slowed down or stopped by applying an electric force. By comparing applied electric force with changes in motion, Millikan was able to determine the electric charge on each drop. After he had measured many drops, he found that the charges on all of them were simple multiples of a single number. This basic unit of charge was the charge on the electron, and the different charges on the oil drops corresponded to those having 2, 3, 4,... extra electrons on them. The charge on the electron is now accepted to be $1.602176565 \times 10^{-19}$ coulomb. For this work Millikan was awarded the Nobel Prize for Physics in 1923.

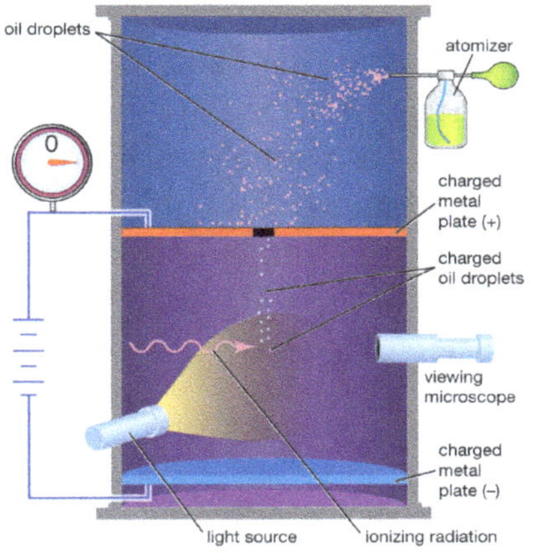

Millikan oil-drop experiment

Between 1909 and 1910 the American physicist Robert Millikan conducted a series of oil-drop experiments. By comparing applied electric force with changes in the motion of the oil drops, he was able to determine the electric charge on each drop. He found that all of the drops had charges that were simple multiples of a single number, the fundamental charge of the electron.

The charge on the proton is equal in magnitude to that on the electron but opposite in sign—that is, the proton has a positive charge. Because opposite electric charges attract each other, there is an attractive force between electrons and protons. This force is what keeps electrons in orbit around the nucleus, something like the way that gravity keeps Earth in orbit around the Sun.

The electron has a mass of about $9.109382911 \times 10^{-28}$ gram. The mass of a proton or neutron is about 1,836 times larger. This explains why the mass of an atom is primarily determined by the mass of the protons and neutrons in the nucleus.

The electron has other intrinsic properties. One of these is called spin. The electron can be pictured as being something like Earth, spinning around an axis of rotation. In fact, most elementary particles have this property. Unlike Earth, however, they exist in the subatomic world and are governed by the laws of quantum mechanics. Therefore, these particles cannot spin in any arbitrary way, but only at certain specific rates. These rates can be $^1/_2$, 1, $^3/_2$, 2,... times a basic unit of rotation. Like protons and neutrons, electrons have spin $^1/_2$.

Particles with half-integer spin are called fermions, for the Italian American physicist Enrico Fermi, who investigated their properties in the first half of the 20th century. Fermions have one important property that will help explain both the way that electrons are arranged in their orbits and the way that protons and neutrons are arranged inside the nucleus. They are subject to the Pauli exclusion principle (named for the Austrian physicist Wolfgang Pauli), which states that no two fermions can occupy the same state—for example, the two electrons in a heliumatom must have different spin directions if they occupy the same orbit.

Because a spinning electron can be thought of as a moving electric charge, electrons can be thought of as tiny electromagnets. This means that, like any other magnet, an electron will respond to the presence of a magnetic field by twisting. (Think of a compass needle pointing north under the influence of Earth's magnetic field.) This fact is usually expressed by saying that electrons have a magnetic moment. In physics, magnetic moment relates the strength of a magnetic field to the torque experienced by a magnetic object. Because of their intrinsic spin, electrons have a magnetic moment given by -9.28×10^{-24} joule per tesla.

Orbits and Energy Levels

Unlike planets orbiting the Sun, electrons cannot be at any arbitrary distance from the nucleus; they can exist only in certain specific locations called allowed orbits. This property, first explained by Danish physicist Niels Bohr in 1913, is another result of quantum mechanics—specifically, the requirement that the angular momentum of an electron in orbit, like everything else in the quantum world, come in discrete bundles called quanta.

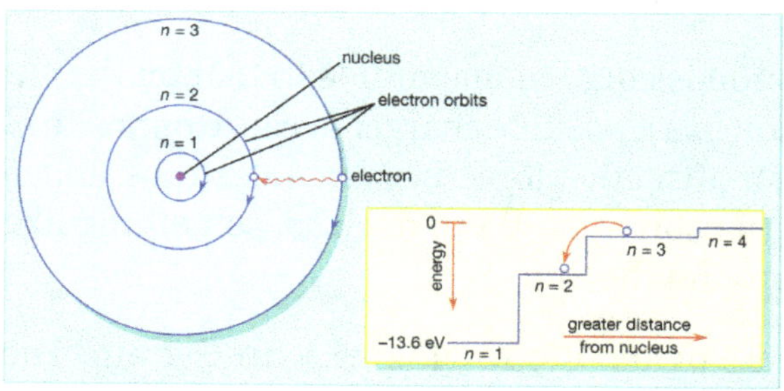

Bohr atom

The electron travels in circular orbits around the nucleus. The orbits have quantized sizes and energies. Energy is emitted from the atom when the electron jumps from one orbit to another closer to the nucleus. Shown here is the first Balmer transition, in which an electron jumps from orbit $n = 3$ to orbit $n = 2$, producing a photon of red light with an energy of 1.89 eV and a wavelength of 656 nanometres.

In the Bohr atom electrons can be found only in allowed orbits, and these allowed orbits are at different energies. The orbits are analogous to a set of stairs in which the gravitational potential energy is different for each step and in which a ball can be found on any step but never in between.

The laws of quantum mechanics describe the process by which electrons can move from one allowed orbit, or energy level, to another. As with many processes in the quantum world, this process is impossible to visualize. An electron disappears from the orbit in which it is located and reappears in its new location without ever appearing any place in between. This process is called a quantum leap or quantum jump, and it has no analog in the macroscopic world.

Because different orbits have different energies, whenever a quantum leap occurs, the energy possessed by the electron will be different after the jump. For example, if an electron jumps from a higher to a lower energy level, the lost energy will have to go somewhere and in fact will be emitted by the atom in a bundle of electromagnetic radiation. This bundle is known as a photon, and this emission of photons with a change of energy levels is the process by which atoms emit light.

In the same way, if energy is added to an atom, an electron can use that energy to make a quantum leap from a lower to a higher orbit. This energy can be supplied in many ways. One common way is for the atom to absorb a photon of just the right frequency. For example, when white light is shone on an atom, it selectively absorbs those frequencies corresponding to the energy differences between allowed orbits.

Each element has a unique set of energy levels, and so the frequencies at which it absorbs and emits light act as a kind of fingerprint, identifying the particular element. This property of atoms has given rise to spectroscopy, a science devoted to identifying atoms and molecules by the kind of radiation they emit or absorb.

This figure of the atom, with electrons moving up and down between allowed orbits, accompanied by the absorption or emission of energy, contains the essential features of the Bohr atomic model, for which Bohr received the Nobel Prize for Physics in 1922. His basic model does not work well in explaining the details of the structure of atoms more complicated than hydrogen, however. This requires the introduction of quantum mechanics. In quantum mechanics each orbiting electron is represented by a mathematical expression known as a wave function—something like a vibrating guitar string laid out along the path of the electron's orbit. These waveforms are called orbitals.

Electron Shells

In the quantum mechanical version of the Bohr atomic model, each of the allowed electron orbits is assigned a quantum number n that runs from 1 (for the orbit closest to the nucleus) to infinity (for orbits very far from the nucleus). All of the orbitals that have the same value of n make up a shell. Inside each shell there may be subshells corresponding to different rates of rotation and orientation of orbitals and the spin directions of the electrons. In general, the farther away from the nucleus a shell is, the more subshells it will have.

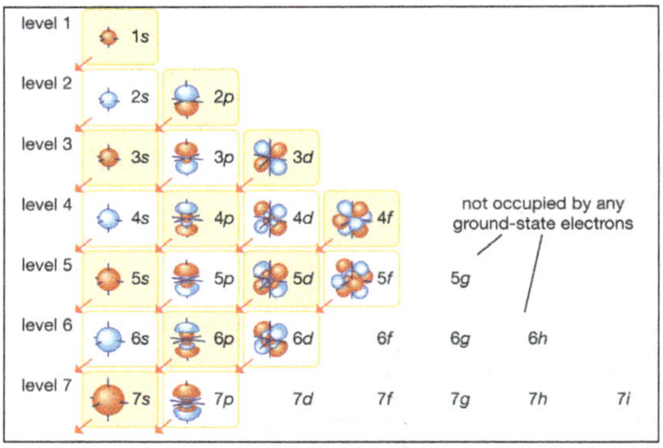

Atomic orbitals

Electrons fill in shell and subshell levels in a semiregular process, as indicated by the arrows above. After filling the first shell level (with just an s subshell), electrons move into the second level s subshell and then into the p subshell, before starting on another shell level. Because of its lower energy state, the 4s orbital fills before the 3d, and similarly for later s orbitals (for example, 6s fills before 4f).

This arrangement of possible orbitals explains a great deal about the chemical properties of different atoms. The easiest way to see this is to imagine building up complex atoms by starting with hydrogen and adding one proton and one electron (along with the appropriate number of neutrons) at a time. In hydrogen the lowest-energy orbit—called the ground state—corresponds to the electron located in the shell closest to the nucleus. There are two possible states for an electron in this shell, corresponding to a clockwise spin and a counterclockwise spin (or, in the jargon of physicists, spin up and spin down).

The next most-complex atom is helium, which has two protons in its nucleus and two orbiting electrons. These electrons fill the two available states in the lowest shell, producing what is called a filled shell. The next atom is lithium, with three electrons. Because the closest shell is filled, the third electron goes into the next higher shell. This shell has spaces for eight electrons, so that it takes an atom with 10 electrons (neon) to fill the first two levels. The next atom after neon, sodium, has 11 electrons, so that one electron goes into the next highest shell.

In the progression thus far, three atoms—hydrogen, lithium, and sodium—have one electron in the outermost shell. As stated above, it is these outermost electrons that determine the chemical properties of an atom. Therefore, these three elements should have similar properties, as indeed they do. For this reason, they appear in the same column of the periodic table of the elements, and the same principle determines the position of every element in that table. The outermost shell of electrons—called the valence shell—determines the chemical behaviour of an atom, and the number of electrons in this shell depends on how many are left over after all the interior shells are filled.

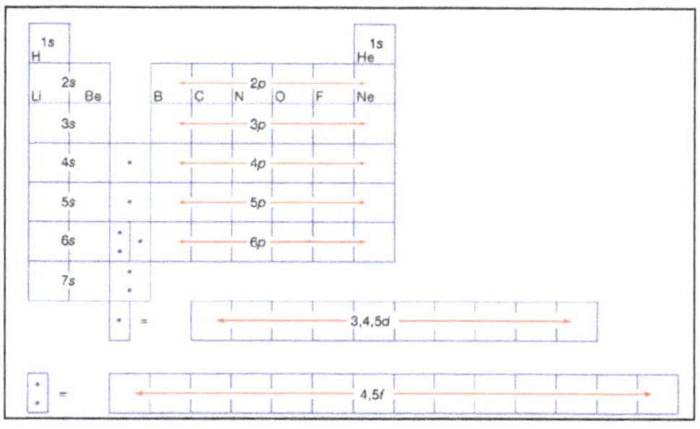

Periodic table of the elements showing the valence shells.

Atomic Bonds

Once the way atoms are put together is understood, the question of how they interact with each other can be addressed—in particular, how they form bonds to create molecules and macroscopic materials. There are three basic ways that the outer electrons of atoms can form bonds:

1. Electrons can be transferred from one atom to another.
2. Electrons can be shared between neighbouring atoms.
3. Electrons can be shared with all atoms in a material.

The first way gives rise to what is called an ionic bond. Consider as an example an atom of sodium, which has one electron in its outermost orbit, coming near an atom of chlorine, which has seven. Because it takes eight electrons to fill the outermost shell of these atoms, the chlorine atom can be thought of as missing one electron. The sodium atom donates its single valence electron to fill the hole in the chlorine shell, forming a sodium chloride system at a lower total energy level.

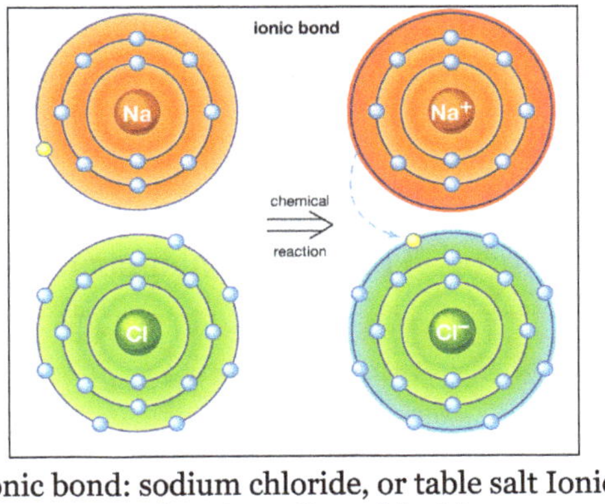

Ionic bond: sodium chloride, or table salt Ionic bonding in sodium chloride.

An atom of sodium (Na) donates one of its electrons to an atom of chlorine (Cl) in a chemical reaction, and the resulting positive ion (Na^+) and negative ion (Cl^-) form a stable ionic compound (sodium chloride; common table salt) based on this ionic bond.

An atom that has more or fewer electrons in orbit than protons in its nucleus is called an ion. Once the electron from its valence shell has been transferred, the sodium atom will be missing an electron; it therefore will have a positive charge and become a sodium ion. Simultaneously, the chlorine atom, having gained an extra electron, will take on a negative charge and become a chlorine ion. The electrical force between these two oppositely charged ions is attractive and locks them together. The resulting sodium chloride compound is a cubic crystal, commonly known as ordinary table salt.

The second bonding strategy listed above is described by quantum mechanics. When two atoms come near each other, they can share a pair of outermost electrons (think of the atoms as tossing the electrons back and forth between them) to form a covalent bond. Covalent bonds are particularly common in organic materials, where molecules often contain long chains of carbon atoms (which have four electrons in their valence shells).

Finally, in some materials each atom gives up an outer electron that then floats freely—in essence, the electron is shared by all of the atoms within the material. The electrons form a kind of sea in which the positive ions float like marbles in molasses. This is called the metallic bond and, as the name implies, it is what holds metals together.

There are also ways for atoms and molecules to bond without actually exchanging or sharing electrons. In many molecules the internal forces are such that the electrons tend to cluster at one end of the molecule, leaving the other end with a positive charge. Overall, the molecule has no net electric charge—it is just that the positive and negative charges are found at different places. For example, in water (H_2O) the electrons tend to spend most of their time near the oxygen atom, leaving the region of the hydrogen atoms with a positive charge. Molecules whose charges are arranged in this way are called polar molecules. An atom or ion approaching a polar molecule from its negative side, for example, will experience a stronger negative electric force than the more-distant positive electric force. This is why many substances dissolve in water: the polar water molecule can pull ions out of materials by exerting electric forces. A special case of polar forces occurs in what is called the hydrogen bond. In many situations, when hydrogen forms a covalent bond with another atom, electrons move toward that atom, and the hydrogen acquires a slight positive charge. The hydrogen, in turn, attracts another atom, thereby forming a kind of bridge between the two. Many important molecules, including DNA, depend on hydrogen bonds for their structure.

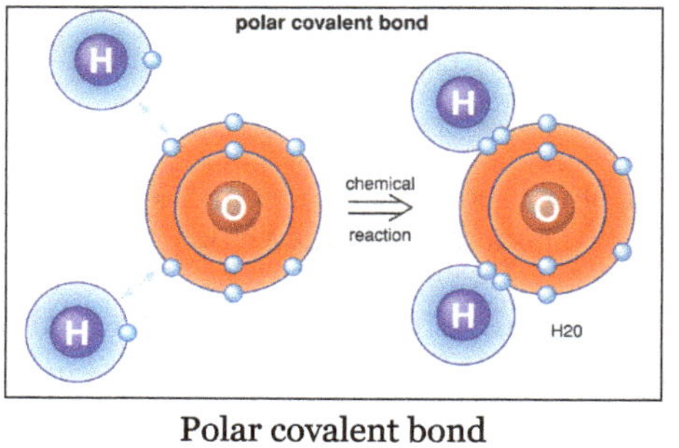

Polar covalent bond

In polar covalent bonds, such as that between hydrogen and oxygen atoms, the electrons are not transferred from one atom to the other as they are in an ionic bond. Instead, some outer electrons merely spend more time in the vicinity of the other atom. The effect of this orbital distortion is to induce regional net charges that hold the atoms together, such as in water molecules.

Finally, there is a way for a weak bond to form between two electrically neutral atoms. Dutch physicist Johannes van der Waals first theorized a mechanism for such a bond in 1873, and it is now known as van der Waals forces. When two atoms approach each other, their electron clouds exert repulsive forces on each other, so that the atoms become polarized. In such situations, it is possible that the electrical attraction between the nucleus of one atom and the electrons of the other will overcome the repulsive forces between the electrons, and a weak bond will form. One example of this force can be seen in ordinary graphite pencil lead. In this material, carbon atoms are held together in sheets by strong covalent bonds, but the sheets are held together only by van der Waals forces. When a pencil is drawn across paper, the van der Waals forces break, and sheets of carbon slough off. This is what creates the dark pencil streak.

Conductors and Insulators

The way that atoms bond together affects the electrical properties of the materials they form. For example, in materials held together by the metallic bond, electrons float loosely between the metal ions. These electrons will be free to move if an electrical force is applied. For example, if a copper wire is attached across the poles of a battery, the electrons will flow inside the wire. Thus, an electric current flows, and the copper is said to be a conductor.

The flow of electrons inside a conductor is not quite so simple, though. A free electron will be accelerated for a while but will then collide with an ion. In the collision process, some of the energy acquired by the electron will be transferred to the ion. As a result, the ion will move faster, and an observer will notice the wire's temperature rise. This conversion of electrical energy from the motion of the electrons to heat energy is called electrical resistance. In a material of high resistance, the wire heats up quickly as electric current flows. In a material of low resistance, such as copper wire, most of the energy remains with the moving electrons, so the material is good at moving electrical energy from one point to another. Its excellent conducting property, together with its relatively low cost, is why copper is commonly used in electrical wiring.

The exact opposite situation obtains in materials, such as plastics and ceramics, in which the electrons are all locked into ionic or covalent bonds. When these kinds of materials are placed between the poles of a battery, no current flows—there are simply no electrons free to move. Such materials are called insulators.

Magnetic Properties

The magnetic properties of materials are also related to the behaviour of electrons in atoms. An electron in orbit can be thought of as a miniature loop of electric current. According to the laws of electromagnetism, such a loop will create a magnetic field. Each electron in orbit around a nucleus produces its own magnetic field, and the sum of these fields, together with the intrinsic fields of the electrons and the nucleus, determines the magnetic field of the atom. Unless all of these fields cancel out, the atom can be thought of as a tiny magnet.

In most materials these atomic magnets point in random directions, so that the material itself is not magnetic. In some cases—for instance, when randomly oriented atomic magnets are placed in a strong external magnetic field—they line up, strengthening the external field in the process. This phenomenon is known as paramagnetism. In a few metals, such as iron, the interatomic forces are such that the atomic magnets line up over regions a few thousand atoms across. These regions are called domains. In normal iron the domains are oriented randomly, so the material is not magnetic. If iron is put in a strong magnetic field, however, the domains will line up, and they will stay lined up even after the external field is removed. As a result, the piece of iron will acquire a strong magnetic field. This phenomenon is known as ferromagnetism. Permanent magnets are made in this way.

The Nucleus

Nuclear Forces

The primary constituents of the nucleus are the proton and the neutron, which have approximately equal mass and are much more massive than the electron. For reference, the accepted mass of the proton is $1.672621777 \times 10^{-24}$ gram, while that of the neutron is $1.674927351 \times 10^{-24}$ gram. The charge on the proton is equal in magnitude to that on the electron but is opposite in sign, while the neutron has no electrical charge. Both particles have spin $^1/_2$ and are therefore fermions and subject to the Pauli exclusion principle. Both also have intrinsic magnetic fields. The magnetic moment of the proton is $1.410606743 \times 10^{-26}$ joule per tesla, while that of the neutron is $-0.96623647 \times 10^{-26}$ joule per tesla.

It would be incorrect to picture the nucleus as just a collection of protons and neutrons, analogous to a bag of marbles. In fact, much of the effort in physics research during the second half of the 20th century was devoted to studying the various kinds of particles that live out their fleeting lives inside the nucleus. A more-accurate picture of the nucleus would be of a seething cauldron where hundreds of different kinds of particles swarm around the protons and neutrons. It is now believed that these so-called elementary particles are made of still more-elementary objects, which have been given the name of quarks. Modern theories suggest that even the quarks may be made of still more-fundamental entities called strings.

The forces that operate inside the nucleus are a mixture of those familiar from everyday life and those that operate only inside the atom. Two protons, for example, will repel each other because of their identical electrical force but will be attracted to each other by gravitation. Especially at the scale of elementary particles, the gravitational force is many orders of magnitude weaker than other fundamental forces, so it is customarily ignored when talking about the nucleus. Nevertheless, because the nucleus stays together in spite of the repulsive electrical force between protons, there must exist a counterforce—which physicists have named the strong force—operating at short range within the nucleus. The strong force has been a major concern in physics research since its existence was first postulated in the 1930s.

One more force—the weak force—operates inside the nucleus. The weak force is responsible for some of the radioactive decays of nuclei. The four fundamental forces—strong, electromagnetic, weak, and gravitational—are responsible for every process in the universe. One of the important strains in modern theoretical physics is the idea that, although they seem very different, they are different aspects of a single underlying force.

Nuclear Shell Model

Many models describe the way protons and neutrons are arranged inside a nucleus. One of the most successful and simple to understand is the shell model. In this model the protons and neutrons occupy separate systems of shells, analogous to the shells in which electrons are found outside the nucleus. From light to heavy nuclei, the proton and neutron shells are filled (separately) in much the same way as electron shells are filled in an atom.

Like the Bohr atomic model, the nucleus has energy levels that correspond to processes in which protons and neutrons make quantum leaps up and down between their allowed orbits. Because energies in the nucleus are so much greater than those associated with electrons, however, the photons emitted or absorbed in these reactions tend to be in the X-ray or gamma ray portions of the electromagnetic spectrum, rather than the visible light portion.

When a nucleus forms from protons and neutrons, an interesting regularity can be seen: the mass of the nucleus is slightly less than the sum of the masses of the constituent protons and neutrons. This consistent discrepancy is not large—typically only a fraction of a percent—but it is significant. By Albert Einstein's principles of relativity, this small mass deficit can be converted into energy via the equation $E = mc^2$. Thus, in order to break a nucleus into its constituent protons and neutrons, energy must be supplied to make up this mass deficit. The energy corresponding to the mass deficit is called the binding energy of the nucleus, and, as the name suggests, it represents the energy required to tie the nucleus together. The binding energy varies across the periodic table and is at a maximum for iron, which is thus the most stable element.

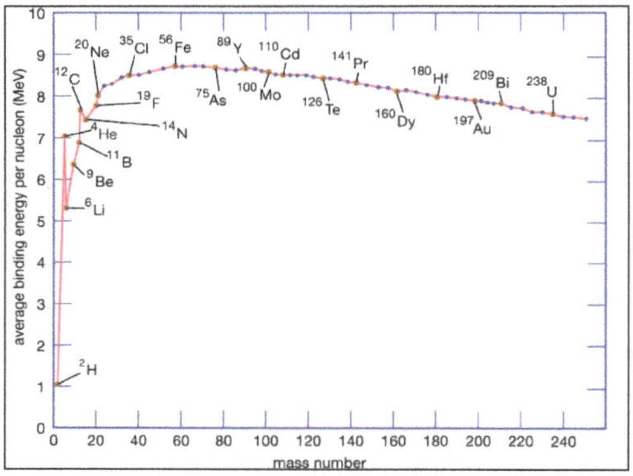

Nuclear binding energies, shown as a function of atomic mass number.

Radioactive Decay

The nuclei of most everyday atoms are stable—that is, they do not change over time. This statement is somewhat misleading, however, because nuclei that are not stable generally do not last long and hence tend not to be part of everyday experience. In fact, most of the known isotopes of nuclei are not stable; instead, they go through a process called radioactive decay, which often changes the identity of the original atom.

In radioactive decay a nucleus will remain unchanged for some unpredictable period and then emit a high-speed particle or photon, after which a different nucleus will have replaced the original. Each unstable isotope decays at a different rate; that is, each has a different probability of decaying within a given period of time. A collection of identical unstable nuclei do not all decay at once. Instead, like popcorn popping in a pan, they will decay individually over a period of time. The time that it takes for half of the original sample to decay is called the half-life of the isotope. Half-lives of known isotopes range from microseconds to billions of years. Uranium-238 (^{238}U) has a half-life of about 4.5 billion years, which is approximately the time that has elapsed since the formation of the solar system. Thus, Earth has about half of the ^{238}U that it had when it was formed.

There are three different types of radioactive decay. In the late 19th century, when radiation was still mysterious, these forms of decay were denoted alpha, beta, and gamma. In alpha decay a nucleus ejects two protons and two neutrons, all locked together in what is called an alpha particle (later discovered to be identical to the nucleus of a normal helium atom). The daughter, or decayed, nucleus will have two fewer protons and two fewer neutrons than the original and hence will be the nucleus of a different chemical element. Once the electrons have rearranged themselves (and the two excess electrons have wandered off), the atom will, in fact, have changed identity.

In beta decay one of the neutrons in the nucleus turns into a proton, a fast-moving electron, and a particle called a neutrino. This emission of fast electrons is called beta radiation. The daughter nucleus has one fewer neutron and one more proton than the original and hence, again, is a different chemical element.

In gamma decay a proton or neutron makes a quantum leap from a higher to a lower orbit, emitting a high-energy photon in the process. In this case the chemical identity of the daughter nucleus is the same as the original.

When a radioactive nucleus decays, it often happens that the daughter nucleus is radioactive as well. This daughter will decay in turn, and the daughter nucleus of that decay may be radioactive as well. Thus, a collection of identical atoms may, over time, be turned into a mixture of many kinds of atoms because of successive decays. Such decays will continue until stable daughter nuclei are produced. This process, called a decay chain, operates everywhere in nature. For example, uranium-238 decays with a half-life of 4.5 billion years into thorium-234, which decays in 24 days into protactinium-234, which also decays. This process continues until it gets to lead-206, which is stable. Dangerous elements such as radium and radon are continually produced in Earth's crust as intermediary steps in decay chains.

Nuclear Energy

It is almost impossible to have lived at any time since the mid-20th century and not be aware that energy can be derived from the atomic nucleus. The basic physical principle behind this fact is that the total mass present after a nuclear reaction is less than before the reaction. This difference in mass, via the equation $E = mc^2$, is converted into what is called nuclear energy.

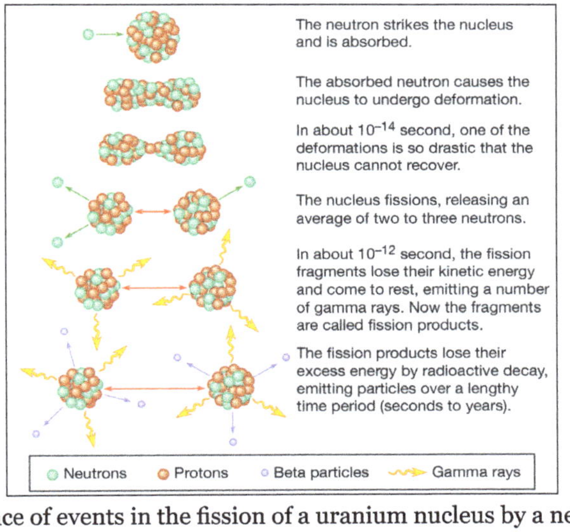

Sequence of events in the fission of a uranium nucleus by a neutron.

There are two types of nuclear processes that can produce energy—nuclear fission and nuclear fusion. In fission a heavy nucleus (such as uranium) is split into a collection of lighter nuclei and fast-moving particles. The energy at the end typically appears in the kinetic energy of the final particles. Nuclear fission is used in nuclear reactors to produce commercial electricity. It depends on the fact that a particular isotope of uranium (^{235}U) behaves in a particular way when it is hit by a neutron. The nucleus breaks apart and emits several particles. Included in the debris of the fission are two or three more free neutrons that can produce fission in other nuclei in a chain reaction. This chain reaction can be controlled and used to heat water into steam, which can then be used to turn turbines in an electrical generator.

Fusion refers to a process in which two or more light nuclei come together to form a heavier nucleus. The most common fusion process in nature is one in which four protons come together to form a helium nucleus (two protons and two neutrons) and some other particles. This is the process by which energy is generated in stars. Scientists have not yet learned to produce a controllable, commercially useful nuclear fusion on Earth, which remains a goal for the future.

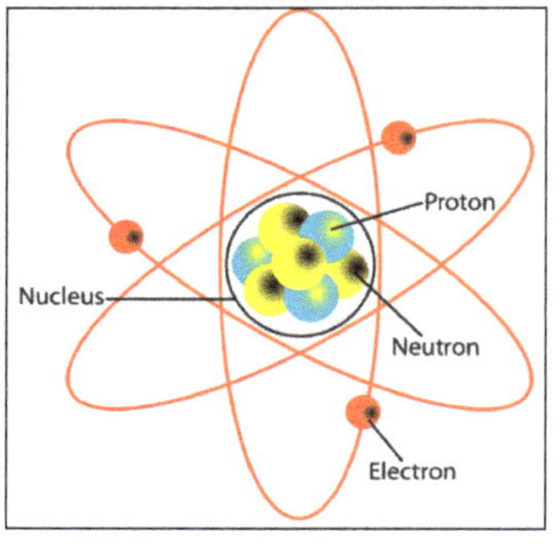

Major parts of an atom

Relative Atomic Mass

The Relative Atomic Mass of an element is the mass of an 'average atom' of that element (taking into account its different isotopes and their relative proportions) compared with the mass of an atom of carbon-12.

It is useful to understand the following in order to be able to explain relative atomic mass: atom, atomic number, element, mass number and isotope.

At the center of each atom is a nucleus that consists of two types of particles called protons and neutrons (and known collectively as nucleons). The number of protons is called the atomic number and determines which element the atom is a particle of (and therefore also its chemical properties). The total number of protons and neutrons in the nucleus of the atom is called the mass number and is determined by the type of atom (i.e. which element it is) and also by the number of neutrons present in that particular atom. In many cases, different atoms of the same element can have different numbers of neutrons in their nuclei (nuclei is the plural word for nucleus; one nucleus - many nuclei). The number of neutrons in an atom of a specific element therefore determines its mass number and so the isotope of which that particular atom is an example.

For example, most atoms of the element phosphorus have a mass number of 31 and so may be referred to as phosphorus-31, however, some atoms of phosphorus have an extra neutron and therefore a mass number of 32. They are therefore atoms of the isotope phosphorus-32.

Although there are three types of particles that form atoms (protons, neutrons and electrons), mass of atoms is mainly due to the numbers of protons and neutrons that make-up the atom, i.e. the mass number of the atom.

- The atomic number of an atom is the same as the atomic number of the element of which that atom is a particle because the atomic number of an atom determines which element it is an atom of.
- The mass number of an atom is not the same as the mass number of an element because atoms of the same element can have different masses due to having different numbers of neutrons (i.e. exist in the form of different isotopes, which therefore have slightly different physical properties e.g. boiling points).

That is the Atomic Number of an element is meaningful, but the Mass Number 'of an element' is not meaningful.

When dealing with elements (macroscopic scale) rather than atoms (individual particles), chemists use Relative Atomic Mass because 'mass number' only applies to specific atoms of one particular isotope of the element.

Calculation of Relative Atomic Mass

In order to calculate the relative atomic mass of a sample of an element, it is necessary to know which isotopes of that element are present in the sample, and in what proportions.

The following calculation of relative atomic mass uses the example of chlorine:

Consider a sample of chlorine that contains the isotopes chlorine-35 and chlorine-37 in the ratio 3:1, meaning that for every one atom of chlorine-37 there are 3 atoms of chlorine-35.

The relative atomic mass of this sample of chlorine may be calculated by taking into account the atomic masses and relative proportions of its isotopes as follows:

	Isotope 1	Isotope 2	
Name	chlorine-35	chlorine-37	
Mass Number	35	37	
Relative Proportion	3	1	
(Mass Number) x (Relative Proportion)	35 x 3 = 105	37 x 1 = 37	
(Mass Number) x (Relative Proportion) of Isotope 1 + (Mass Number) x (Relative Proportion) of Isotope 2			= 105 + 37 = 142
Answer from above divided by the sum of the relative proportions of the isotopes			= 142 / (3+1)
Relative Atomic Mass of this sample			= 142 / 4 = 35.5

Molecules

A molecule is the smallest particle in a chemical element or compound that has the chemical properties of that element or compound. Molecules are made up of atoms that are held together by chemical bonds. These bonds form as a result of the sharing or exchange of electrons among atoms. The atoms of certain elements readily bond with other atoms to form molecules. Examples of such elements are oxygen and chlorine. The atoms of some elements do not easily bond with other atoms. Examples are neon and argon.

Molecules can vary greatly in size and complexity. The element helium is a one-atom molecule. Some molecules consist of two atoms of the same element. For example, O_2 is the oxygen molecule most commonly found in the earth's atmosphere; it has two atoms of oxygen. However, under certain circumstances, oxygen atoms bond into triplets (O_3), forming a molecule known as ozone. Other familiar molecules include water, consisting of two hydrogen atoms and one oxygen atom (H_2O), carbon dioxide, consisting of one carbon atom bonded to two oxygen atoms (CO_2), and sulfuric acid, consisting of two hydrogen atoms, one sulfur atom, and four oxygen atoms (H_2SO_4).

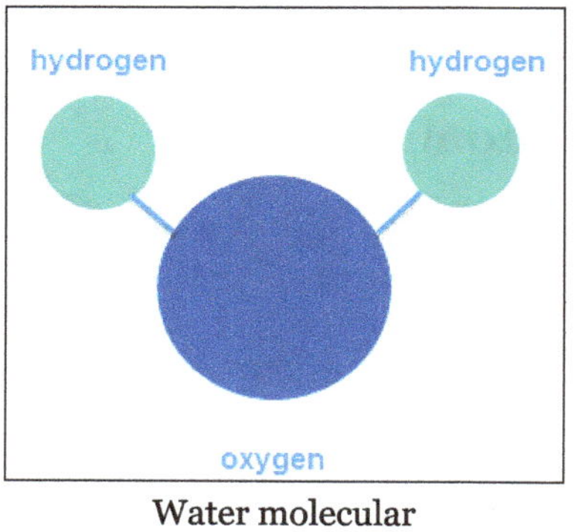

Water molecular

Some molecules, notably certain proteins, contain hundreds or even thousands of atoms that join together in chains that can attain considerable lengths. Liquids containing such molecules sometimes behave strangely. For example, a liquid may continue to flow out of a flask from which some of it has been poured, even after the flask is returned to an upright position.

Molecules are always in motion. In solids and liquids, they are packed tightly together. In a solid, the motion of the molecules can be likened to rapid vibration. In a liquid, the molecules can move freely among each other, in a sort of slithering fashion. In a gas, the density of molecules is generally less than in a liquid or solid of the same chemical compound, and they move even more freely than in a liquid. For a specific compound in a given state (solid, liquid, or gas), the speed of molecular motion increases as the absolute temperature increases.

Molecular Mass and Mole Concept

Molecular mass of a substance is the relative mass of its molecule when compared with the mass of 12C atom considered as 12-units. In simple terms, it points out the number of times; 1 molecule of the concerned substance is weightier than an atom.

- Calculation of Molecular Mass: Always remember that molecular mass is equivalent to the sum of atomic masses for all atoms available in a molecule of the target substance.

Relative Molecular Mass

RMM or Relative molecular mass is the molecular weight of an element or compound. You can call it as the number of times a single molecule of the substance stays heavier than the $1/12^{th}$ mass of carbon atom (12C).

Gram Molecular Mass

Gram molecular mass of a substance is the molecular mass denoted in grams. For example, molecular mass for O_2 = 32 Gram; the relative molecular mass as expressed in terms of grams. Do remember, relative atomic mass stays as a ratio and doesn't have any units. On the other hand, gram molecular mass and gram atomic mass can be denoted in grams.

Concept of Mole

It is vital to note that, the identity of a substance not only contains the type of atoms but also the quantity of each atom type. In today's time, we have access to sophisticated instruments which permit the direct measurement microscopic traits.

But, the same traits were previously derived from the evaluation of macroscopic properties using comparatively simple tools. Such an experimental approach did require the introduction of a fresh unit for defining the quantity of substances, called as the mole.

It is interesting to know that, to date, this unit stays essential for modern chemical science. What is a mole? It is an amount unit that can be seen similar to familiar units such as a pair, gross, dozen etc. To be precise, it offers a specific measure of the count of atoms or molecules present in a sample bulk matter.

Furthermore, a mole can be defined as the quantity of substance comprising the same number of distinct entities (like atoms, ions, and molecules) as the count of atoms in a sample comprising pure 12C weighing accurately 12 g. The mole offers a link between a simply calculated macroscopic property (bulk mass) and a truly significant fundamental property (number of atoms, molecules etc).

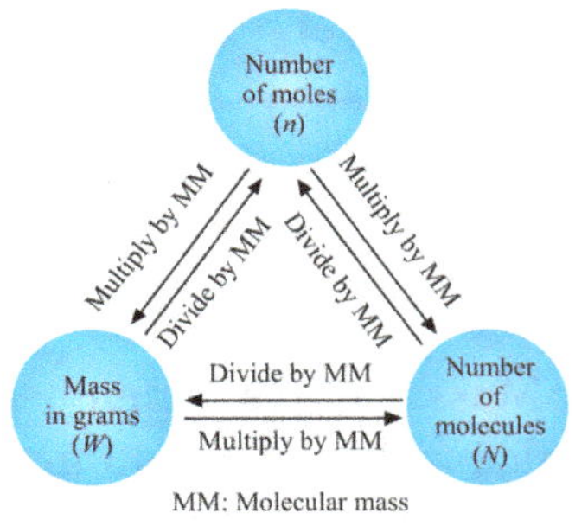

Figure: Molecular mass

The count of units forming a mole has been determined experimentally as 6.02214179 × 1023. This is called as the fundamental constant termed as Avogadro's number (NA) or Avogadro constant. In chemistry, this constant is suitably reported using an explicit unit called per mole.

Atomic Structure and Model

The atomic structure of an element refers to the constitution of its nucleus and the arrangement of the electrons around it. Primarily, the atomic structure of matter is made up of protons, electrons, and neutrons.

The protons and neutrons make up the nucleus of the atom, which is surrounded by the electrons belonging to the atom. The atomic number of an element describes the total number of protons in its nucleus.

Neutral atoms have equal numbers of protons and electrons. However, atoms may gain or lose electrons in order to increase their stability, and the resulting charged entity is called an ion.

Atoms of different elements have different atomic structures because they contain different numbers of protons and electrons. This is the reason for the unique characteristics of different elements.

Atomic Models

In the 18th and 19th centuries, many scientists attempted to explain the structure of the atom with the help of atomic models. Each of these models had their own merits and demerits and were pivotal to the development of the modern atomic model. The most notable contributions to the field were by the scientists John Dalton, J.J. Thomson, Ernest Rutherford, and Niels Bohr.

Dalton's Atomic Model

Matter has been one of the most important subjects of research for the science enthusiasts. Scientists and philosophers have always tried to simplify things and so was the case with the matter. They wanted to know about the fundamental particles that make matter, their properties, structure etc. This led to the formulation of a number of atomic theories.

Democritus was the first man who proposed that matter is made up of particles. He named these particles, 'atomos' meaning indivisible. This was the Democritus' Atomic Theory. Due to the lack of technological setup back then, scientists had very limited information on this theory.

Almost after two thousand years, the works on the simplifying matter was materialized by scientist, John Dalton. In 1808, John Dalton postulated the famous Dalton's Atomic Theory. He published this theory in a paper titled "A New Chemical Philosophy"; indeed the philosophy was new for that era. Let us now look at the postulates of this theory.

Postulates of Dalton's Atomic Theory

- The matter is made up of indivisible particles known as atoms.
- The properties of all the atoms of a given element are the same including mass. This can also be stated as all the atoms of an element have identical mass while the atoms of different elements have different masses.
- Atoms of different elements combine in fixed ratios to form compounds.
- Atoms are neither created nor destroyed. This implies that during chemical reactions, no atoms are created nor destroyed.
- The formation of new products (compounds) results from the rearrangement of existing atoms (reactants).
- Atoms of an element are identical in mass, size and many other chemical or physical properties, but atoms of two-different elements differ in mass, size, and many other chemical or physical properties.

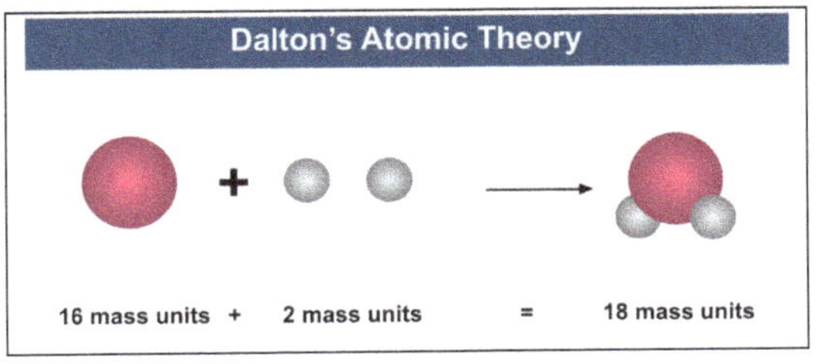

However, this theory is not entirely free of limitations. Let us now look at the drawbacks of this theory.

Drawbacks of Dalton's Atomic Theory of Matter

- It was proved that an atom is not indivisible. As an atom can be subdivided into electrons, protons and neutrons. But remember that atom is the tiniest particle that takes part in a chemical reaction.
- According to Dalton Atomic Theory, atoms of an element are identical in mass, size and many other chemical or physical properties. But, practically we observe that atoms of several elements differ in their densities and masses. These atoms with the different masses are known as isotopes. For example, Chlorine (Cl) has 2 isotopes with the mass numbers of 35 and 37.
- Also, according to Dalton Atomic Theory, atoms of two-different elements differ in mass, size and many other chemical or physical properties. However, this is not correct for all situations. For example, Argon (Ar) and Calcium (Ca) atoms, each have an atomic mass of 40 amu. These atoms with similar atomic masses are isobars.
- According to Dalton Atomic Theory, when atoms of different elements (atoms of two or more elements) combine in simple whole number ratios, we get chemical compounds. But this is not true in case of complex organic compounds.
- Dalton Atomic Theory fails to explain the existence of allotropes. This implies that the Dalton atomic theory fails to explain the differences in properties of charcoal, graphite, and diamond (allotropes of carbon).
- Dalton's Atomic Theory also suggested that an atom is the smallest part of an atom that can take part in a chemical reaction. Some postulates of this theory remain valid even in today's modern chemical thoughts. The atomic structure model proposed by indeed proves to be a significant, stepping stone in chemistry. It forms the base for modern atomic theories and quantum mechanics.

Thomson Atomic Model

Thomson atomic model was proposed by William Thomson in the year 1900. This model explained the description of an inner structure of the atom theoretically. It was strongly supported by Sir Joseph Thomson, who had discovered the electron earlier.

During cathode ray tube experiment, a negatively charged particle was discovered by J.J. Thomson. This experiment took place in the year 1897. Cathode ray tube is a vacuum tube. The negative particle was called an electron.

Thomson assumed that an electron is two thousand times lighter than a proton and believed that an atom is made up of thousands of electrons. In this atomic structure model, he considered atoms surrounded by a cloud having positive as well as negative charges. The demonstration of the ionization of air by X-ray was also done by him together with Rutherford. They were the first to demonstrate it. Thomson's model of an atom is similar to a plum pudding.

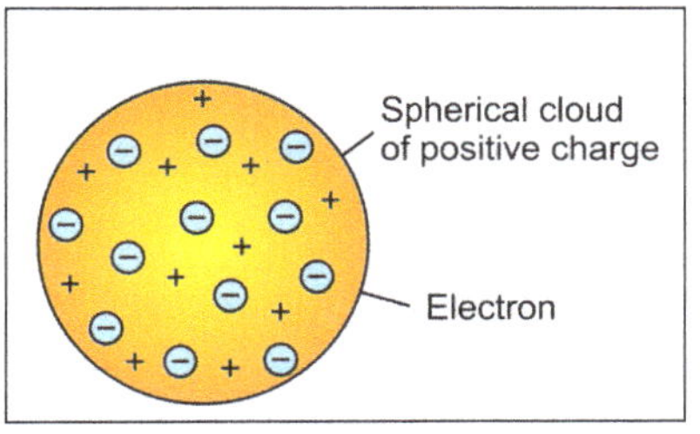

Postulates of Thomson's Atomic Model

Postulate 1: An atom consists of a positively charged sphere with electrons embedded in it.

Postulate 2: An atom as a whole is electrically neutral because the negative and positive charges are equal in magnitude.

Thomson atomic model is compared to watermelon. Where he considered:

- Watermelon seeds as negatively charged particles.
- The red part of the watermelon as positively charged.

Limitations of Thomson's Atomic Model

- It failed to explain the stability of an atom because his model of atom failed to explain how a positive charge holds the negatively charged electrons in an atom. Therefore, This theory also failed to account for the position of the nucleus in an atom.
- Thomson's model failed to explain the scattering of alpha particles by thin metal foils.
- No experimental evidence in its support.

Although Thomson's model was not an accurate model to account for the atomic structure, it proved to be the base for the development of other atomic models.

Rutherford Atomic Model

Rutherford proposed that an atom is composed of empty space mostly with electrons orbiting in a set, predictable paths around fixed, positively charged nucleus.

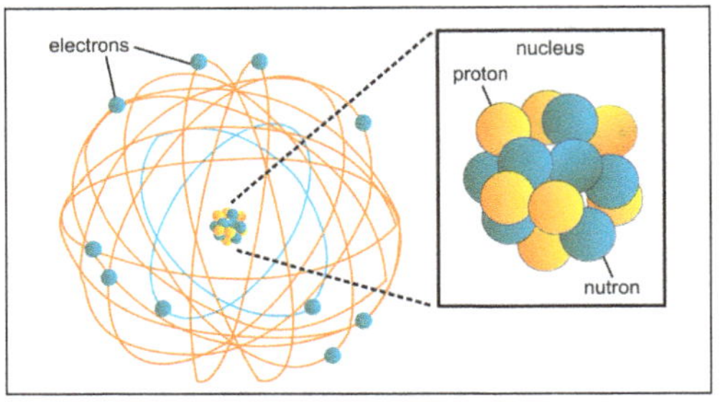

Rutherford's Atomic Model

The concept of atom dates back to 400 BCE when Greek philosopher Democritus first conceived the idea. However, it was not until 1803 John Dalton proposed again the idea of the atom. But at that point of time, atoms were considered indivisible. This idea of an atom as indivisible particles continued until the year 1897 when British Physicist J.J. Thomson discovered negatively charged particles which were later named electrons.

He proposed a model on the basis of that where he explained electrons were embedded uniformly in a positively charged matrix. The model was named plum pudding model. However, J.J. Thomson's plum pudding model had some limitations. It failed to explain certain experimental results related to the atomic structure of elements.

A British Physicist "Ernest Rutherford" proposed a model of the atomic structure known as Rutherford's Model of Atoms. He conducted an experiment where he bombarded α-particles in a thin sheet of gold. In this experiment, he studied the trajectory of the α-particles after interaction with the thin sheet of gold.

Rutherford Atomic Model Experiment

In Rutherford's experiment, he bombarded high energy streams of α-particles on a thin gold foil of 100 nm thickness. The streams of α-particles were directed from a radioactive source. He conducted the experiment to study the deflection produced in the trajectory of α-particles after interaction with the thin sheet of gold. To study the deflection, he placed a screen made up of zinc sulfide around the gold foil. The observations made by Rutherford contradicted the plum pudding model given by J.J. Thomson.

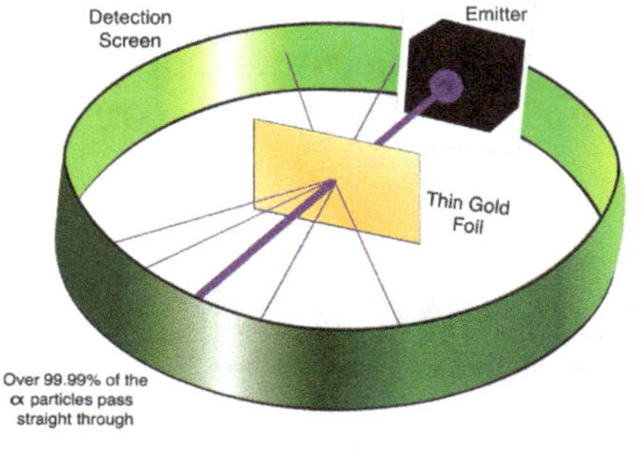

Rutherford's Gold Foil Experiment

Observations of Rutherford Model Experiment

On the basis of the observations made during the experiment, Rutherford concluded that,

- Major space in an atom is empty – A large fraction of α-particles passed through the gold sheet without getting deflected. Therefore, the major part of an atom must be empty.
- The positive charge in an atom is not distributed uniformly and it is concentrated in a very small volume – Few α-particles when bombarded were deflected by the gold sheet. They were deflected minutely and at very small angles. Therefore he made the above conclusion.
- Very few α-particles had deflected at large angles or deflected back. Moreover, very few particles had deflected at 180°. Therefore, he concluded that the positively charged particles covered a small volume of an atom in comparison to the total volume of an atom.

Postulates of Rutherford Atomic Model Based on Observations and Conclusions

- An atom is composed of positively charged particles. Majority of the mass of an atom was concentrated in a very small region. This region of the atom was called as the nucleus of an atom. It was found out later that the very small and dense nucleus of an atom is composed of neutrons and protons.
- Atoms nucleus is surrounded by negatively charged particles called electrons. The electrons revolve around the nucleus in a fixed circular path at very high speed. These fixed circular paths were termed as "orbits."
- An atom has no net charge or they are electrically neutral because electrons are negatively charged and the densely concentrated nucleus is positively charged. A strong electrostatic force of attractions holds together the nucleus and electrons.
- The size of the nucleus of an atom is very small in comparison to the total size of an atom.

Drawbacks of Rutherford Model

Rutherford's experiment was unable to explain certain things. They are:

- Rutherford's model was unable to explain the stability of an atom. According to Rutherford's postulate, electrons revolve at a very high speed around a nucleus of an atom in a fixed orbit. However, Maxwell explained accelerated charged particles release electromagnetic radiations. Therefore, electrons revolving around the nucleus will release electromagnetic radiation.

- The electromagnetic radiation will have energy from the electronic motion as a result of which the orbits will gradually shrink. Finally, the orbits will shrink and collapse in the nucleus of an atom. According to the calculations, if Maxwell's explanation is followed Rutherford's model will collapse with 10-8seconds. Therefore, Rutherford atomic model was not following Maxwell's theory and it was unable to explain an atom's stability.
- Rutherford's theory was incomplete because it did not mention anything about the arrangement of electrons in the orbit. This was one of the major drawbacks of Rutherford atomic model.

Bohr's Atomic Theory

Thomson's atomic model and Rutherford's atomic model failed to answer many questions related to the energy of an atom and its stability. In the year 1913, Niels Bohr proposed an atomic structure model, describing an atom as a small, positively charged nucleus surrounded by electrons that travel in circular orbits around the positively charged nucleus as planets around the sun in our solar system, with attraction provided by electrostatic forces, popularly known as Bohr's atomic model. It was basically an improved version of Rutherford's atomic model overcoming its limitations. On most of the points, he is in agreement with him, like concepts of nucleus and electrons orbiting it. Salient features of Bohr's atomic model are:

- Electrons revolve around the nucleus in stable orbits without emission of radiant energy. Each orbit has a definite energy and is called an energy shell or energy level.
- An orbit or energy level is designated as K, L, M, N shells. When the electron is in the lowest energy level, it is said to be in the ground state.
- An electron emits or absorbs energy when it jumps from one orbit or energy level to another. When it jumps from a higher energy level to lower energy level it emits energy while it absorbs energy when it jumps from a lower energy level to a higher energy level.
- The energy absorbed or emitted is equal to the difference between the energies of the two energy levels (E_1, E_2) and is determined by Plank's equation.

  $\Delta E = E_2 - E_1 = hv$

  Where,

  ΔE = energy absorbed or emitted

  h= Plank's constant

  v= frequency of electromagnetic radiation emitted or absorbed.

- The angular momentum of an electron revolving in energy shells is given by:

  $m_e vr = nh/2\pi$

  Where,

  n= number of corresponding energy shell; 1, 2, 3 .....

  m_e= mass of the electron

  v= velocity

  r=radius

  h= Plank's constant.

Limitations of Bohr Atomic Model Theory

- It violates the Heisenberg Uncertainty Principle. The Bohr atomic model theory considers electrons to have both a known radius and orbit i.e. known position and momentum at the same time, which is impossible according to Heisenberg.
- The Bohr atomic model theory made correct predictions for smaller sized atoms like hydrogen, but poor spectral predictions are obtained when larger atoms are considered.
- It failed to explain the Zeeman effect when the spectral line is split into several components in the presence of a magnetic field.
- It failed to explain the Stark effect when the spectral line gets split up into fine lines in the presence of an electric field.

Bohr Model of Hydrogen

The simplest example of the Bohr Model is for the hydrogen atom ($Z = 1$) or for a hydrogen-like ion ($Z > 1$), in which a negatively-charged electron orbits a small positively-charged nucleus. Electromagnetic energy will be absorbed or emitted if an electron moves from one orbit to another. Only certain electron orbits are permitted. The radius of the possible orbits increases as n^2, where n is the principal quantum number. The $3 \rightarrow 2$ transition produces the first line of the Balmer series. For hydrogen ($Z = 1$) this produces a photon having wavelength 656 nm (red light).

Bohr Model for Heavier Atoms

Heavier atoms contain more protons in the nucleus than the hydrogen atom. More electrons were required to cancel out the positive charge of all of these protons. Bohr believed each electron orbit could only hold a set number of electrons. Once the level was full, additional electrons would be bumped up to the next level. Thus, the Bohr model for heavier atoms described electron shells. The model explained some of the atomic properties of heavier atoms, which had never been reproduced before. For example, the shell model explained why atoms got smaller moving across a period (row) of the periodic table, even though they had more protons and electrons. It also explained why the noble gases were inert and why atoms on the left side of the periodic table attract electrons, while those on the right side lose them. However, the model assumed electrons in the shells didn't interact with each other and couldn't explain why electrons seemed to stack in an irregular manner.

Refinements and Improvements to the Bohr Model

The most prominent refinement to the Bohr model was the Sommerfeld model, which is sometimes called the Bohr-Sommerfeld model. In this model, electrons travel in elliptical orbits around the nucleus rather than in circular orbits. The Sommerfeld model was better at explaining atomic spectral effects, such the Stark effect in spectral line splitting. However, the model couldn't accommodate the magnetic quantum number.

Ultimately, the Bohr model and models based upon it were replaced Wolfgang Pauli's model based on quantum mechanics in 1925. That model was improved to produce the modern model, introduced by Erwin Schrodinger in 1926. Today, the behavior of the hydrogen atom is explained using wave mechanics to describe atomic orbitals.

Periodic Law

The Periodic Law states that the physical and chemical properties of the elements recur in a systematic and predictable way when the elements are arranged in order of increasing atomic number. Many of the properties recur at intervals. When the elements are arranged correctly, the trends in element properties become apparent and can be used to make predictions about unknown or unfamiliar elements, simply based on their placement on the table.

Importance of Periodic Law

Periodic Law is considered to be one of the most important concepts in chemistry. Every chemist makes use of Periodic Law, whether consciously or not, when dealing with the chemical elements, their properties, and their chemical reactions. Periodic Law led to the development of the modern periodic table.

Properties Affected by Periodic Law

The key properties that follow trends according to Periodic Law are atomic radius, ionic radius, ionization energy, electronegativity, and electron affinity.

Atomic and ionic radius are a measure of the size of a single atom or ion. While atomic and ionic radius are different from each other, they follow the same general trend. The radius increases moving down an element group and generally decreases moving left to right across a period or row.

Ionization energy is a measure of how easy it is to remove an electron from an atom or ion. This value decreases moving down a group and increases moving left to right across a period.

Electron affinity is how easily an atom accepts an electron. Using Periodic Law, it becomes apparent the alkaline earth elements have a low electron affinity. In contrast, the halogens readily accept electrons to fill their electron subshells and have high electron affinities. The noble gas elements have practically zero electron affinity because they have full valence electron subshells.

Electronegativity is related to electron affinity. It reflects how easily an atom of an element attracts electrons to form a chemical bond. Both electron affinity and electronegativity tend to decrease moving down a group and increase moving across a period. Electropositivity is another trend governed by Periodic Law. Electropositive elements have low electronegativities (e.g., cesium, francium).

In addition to these properties, there are other characteristics associated with Periodic Law, which may be considered properties of element groups. For example, all of the elements in group I (alkali metals) are shiny, carry a +1 oxidation state, react with water, and occur in compounds rather than as free elements.

Pauli Exclusion Principle

No two electrons can have the same set of quantum numbers. That is, no two electrons can be in the same state.

- The Pauli exclusion principle says that every electron must be in its own unique state. In other words, no electrons in an atom are permitted to have an identical set of quantum numbers.
- The Pauli exclusion principle sits at the heart of chemistry, helping to explain the electron arrangements in atoms and molecules, and helping to rationalize patterns in the periodic table.
- In chemistry the Pauli exclusion principle is applied solely to electrons, which we are about to discuss.
- Wolfgang Pauli received the 1945 Nobel Prize in Physics for his discovery as it applied to electrons.

Four Quantum Numbers

Every electron in an atom can be defined completely by four quantum numbers:

- n: the principal quantum number;
- l: the orbital angular momentum quantum number;
- m_l: the magnetic quantum number;
- m_s: the spin quantum number.

Example of the Pauli Exclusion Principle

Consider argon's electron configuration,

$1s^2\ 2s^2\ 2p^6\ 3s^2\ 3p^6$

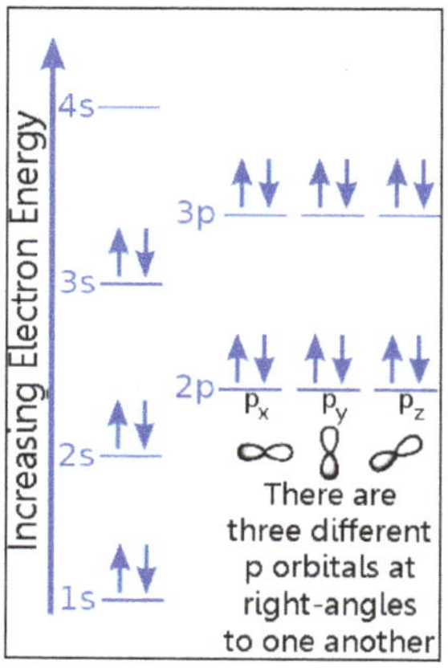

The exclusion principle asserts that every electron in an argon atom is in a unique state:

- The 1s level can accommodate two electrons with identical n, l, and m_l quantum numbers. Argon's pair of electrons in the 1s orbital satisfy the exclusion principle because they have opposite spins, meaning they have different spin quantum numbers, m_s. One spin is +½, the other is -½. (Instead of saying +½ or -½ often the electrons are said to be spin-up ↑or spin-down ↓.)

- The 2s level electrons have a different principal quantum number to those in the 1s orbital. The pair of 2s electrons differ from each other because they have opposite spins.

- The 2p level electrons have a different orbital angular momentum number from those in the s orbitals, hence the letter p rather than s. There are three p orbitals of equal energy, the p_x, p_y and p_z. These orbitals are different from one another because they have different orientations in space. Each of the p_x, p_y and p_z orbitals can accommodate a pair of electrons with opposite spins.

- The 3s level rises to a higher principal quantum number; this orbital accommodates an electron pair with opposite spins.
- The 3p level's description is similar to that for 2p, but the principal quantum number is higher: 3p lies at a higher energy than 2p.

Aufbau Principle

The Aufbau principle dictates the manner in which electrons are filled in the atomic orbitals of an atom in its ground state. It states that electrons are filled into atomic orbitals in the increasing order of orbital energy level. According to the Aufbau principle, the available atomic orbitals with the lowest energy levels are occupied before those with higher energy levels.

The word 'Aufbau' has German roots and can be roughly translated as 'construct' or 'build up'. A diagram illustrating the order in which atomic orbitals are filled is provided below. Here, 'n' refers to the principal quantum number and 'l' is the azimuthal quantum number.

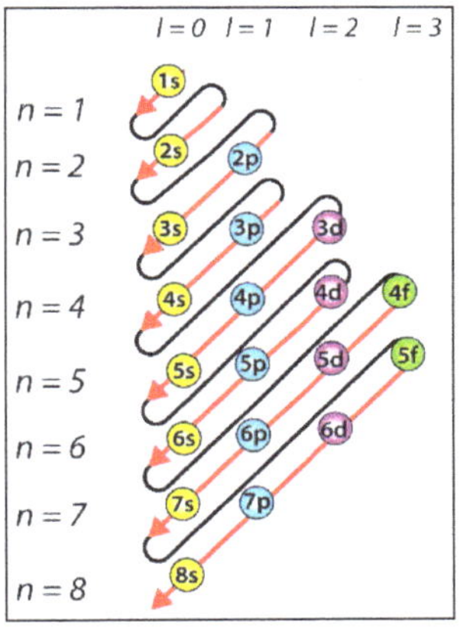

The Aufbau principle can be used to understand the location of electrons in an atom and their corresponding energy levels. For example, carbon has 6 electrons and its electronic configuration is $1s^2 2s^2 2p^2$.

It is important to note that each orbital can hold a maximum of two electrons (as per the Pauli exclusion principle). Also, the manner in which electrons are filled into orbitals in a single subshell must follow Hund's rule, i.e. every orbital in a given subshell must be singly occupied by electrons before any two electrons pair up in an orbital.

Salient Features of the Aufbau Principle

- According to the Aufbau principle, electrons first occupy those orbitals whose energy is the lowest. This implies that the electrons enter the orbitals having higher energies only when orbitals with lower energies have been completely filled.

- The order in which the energy of orbitals increases can be determined with the help of the (n+l) rule, where the sum of the principal and azimuthal quantum numbers determines the energy level of the orbital.
- Lower (n+l) values correspond to lower orbital energies. If two orbitals share equal (n+l) values, the orbital with the lower n value is said to have lower energy associated with it.
- The order in which the orbitals are filled with electrons is: 1s, 2s, 2p, 3s, 3p, 4s, 3d, 4p, 5s, 4d, 5p, 6s, 4f, 5d, 6p, 7s, 5f, 6d, 7p, and so on.

Exceptions

The electron configuration of chromium is $[Ar]3d^5 4s^1$ and not $[Ar]3d^4 4s^2$ (as suggested by the Aufbau principle). This exception is attributed to several factors such as the increased stability provided by half-filled subshells and the relatively low energy gap between the 3d and the 4s subshells.

The energy gap between the different subshells is illustrated below.

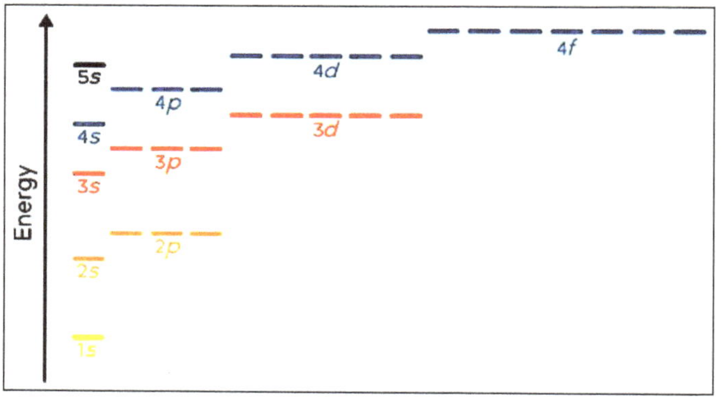

Half filled subshells feature lower electron-electron repulsions in the orbitals, thereby increasing the stability. Similarly, completely filled subshells also increase the stability of the atom. Therefore, the electron configurations of some atoms disobey the Aufbau principle (depending on the energy gap between the orbitals).

For example, copper is another exception to this principle with an electronic configuration corresponding to $[Ar]3d^{10}4s^1$. This can be explained by the stability provided by a completely filled 3d subshell.

Electronic Configuration using the Aufbau Principle

Writing the Electron Configuration of Sulfur

- The atomic number of sulfur is 16, implying that it holds a total of 16 electrons.

- As per the Aufbau principle, two of these electrons are present in the 1s subshell, eight of them are present in the 2s and 2p subshell, and the remaining are distributed into the 3s and 3p subshells.
- Therefore, the electron configuration of sulfur can be written as $1s^2 2s^2 2p^2 3s^2 3p^6$.

Writing the Electron Configuration of Nitrogen

- The element nitrogen has 7 electrons (since its atomic number is 7).
- The electrons are filled into the 1s, 2s, and 2p orbitals.
- The electron configuration of nitrogen can be written as $1s^2 2s^2 2p^3$

Hund's Rule

Aufbau principle tells us that the lowest energy orbitals get filled by electrons first. After the lower energy orbitals are filled, the electrons move on to higher energy orbitals. The problem with this rule is that it does not tell about the three 2p orbitals and the order that they will be filled in. According to Hund's rule:

- Before the double occupation of any orbital, every orbital in the sub level is singly occupied.
- For the maximization of total spin, all electrons in a single occupancy orbital have the same spin.

An electron will not pair with another electron in a half-filled orbital as it has the ability to fill all its orbitals with similar energy. A large number of unpaired electrons are present in atoms which are at the ground state. If two electrons come in contact they would show the same behaviour as two magnets do. The electrons first try to get as far away from each other as possible before they have to pair up.

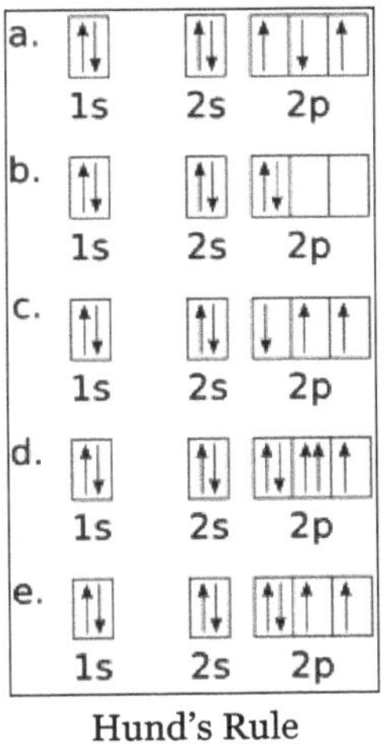

Hund's Rule

State Hund's Rule

It states that:

1. In a sublevel, each orbital is singly occupied before it is doubly occupied.
2. The electrons present in singly occupied orbitals possess identical spin.

Explanation of Hund's Rule

The electrons enter an empty orbital before pairing up. The electrons repel each other as they are negatively charged. The electrons do not share orbitals to reduce repulsion.

When we consider the second rule, the spins of unpaired electrons in singly occupied orbitals is the same. The initial electrons spin in the sub level decides what the spin of the other electrons would be. For instance, a carbon atom's electron configuration would be $1s^2 2s^2 2p^2$. The same orbital will be occupied by the two 2s electrons although different orbitals will be occupied by the two 2p electrons in reference to Hund's rule.

Electron Configuration and its Purpose

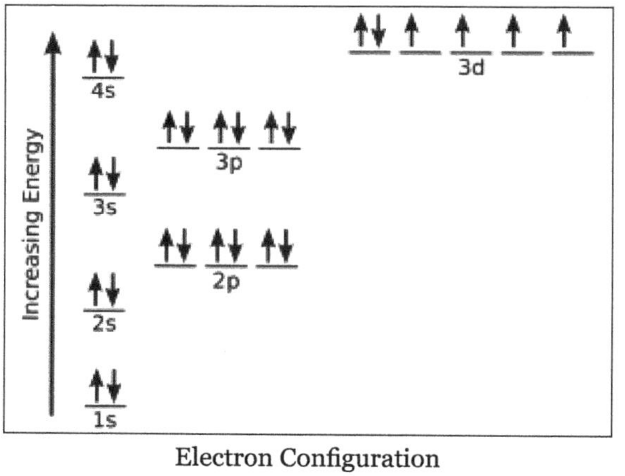

Electron Configuration

The above image helps in understanding the electronic configuration and its purpose. The valence shells of two atoms that come in contact with each other will interact first. When valence shells are not full then the atom is least stable. The chemical characteristics of an element are largely dependent on the valence electrons. Similar chemical characteristics can be seen in elements that have similar valence numbers.

The stability can also be predicted by the electron configuration. When all the orbitals of an atom are full it is most stable. The orbitals that have full energy level are the most stable, for example, noble gases. These type of elements do not react with other elements.

Hund's Rule of Maximum Multiplicity

The rule states that, for a stated electron configuration, the greatest value of spin multiplicity has the lowest energy term. It says if two or more than two orbitals having the same amount of energy are unoccupied then the electrons will start occupying them individually before they fill them in pairs. It is a rule which depends on the observation of atomic spectra, that is helpful in predicting the ground state of a molecule or an atom with one or more than one open electronic shells. This rule was discovered in the year 1925 by Friedrich Hund.

Uses of Hund's Rule

It has wide applications in the following – It is majorly used in atomic chemistry, quantum chemistry, and spectroscopy, etc.

Emission and Absorption Spectra

Spectrum

Ordinary white light consists of waves of all wavelengths in the visible range. This is why, when white light passes through a prism, a series of coloured bands are seen called spectrum. This spectrum of white light ranges from violet at 7.5×10^{14} Hz to red at 4×10^{14} Hz.

Since the colours merge into each other i.e. violet merges into blue, blue into green and so on, we call it a continuous spectrum. When this light passes through an object or medium, the wave with the shortest wavelength (violet) deviates the most than the one with the longest wavelength (red).

The interaction of electromagnetic radiation with matter causes the atoms and molecules to absorb energy and go to a higher energy state. Since this state is unstable, they need to emit radiations to return to their normal states. This gives rise to emission and absorption spectra. Let's learn about them in detail.

Emission and Absorption Spectra

The emission spectrum is the spectrum of radiation emitted by a substance that has absorbed energy. Atoms, molecules, and ions that have absorbed radiation are called 'excited'. The absorption spectrum is the opposite of the emission spectrum.

It is the spectrum formed by electromagnetic radiation that has passed through a medium, in which radiation of some frequencies is absorbed. Spectroscopy is the study of emission and absorption spectra.

Line Spectrum

Unlike visible light which shows a continuous spectrum of all wavelengths, the emission spectra of atoms in the gas phase emit light only at specific wavelengths with dark spaces between them. This is called line spectra or atomic spectra since the emitted radiation is identified by bright lines in the spectra.

Each element has its own unique line emission spectrum. Did you know that just the way fingerprints are used to identify people, the characteristic lines in an atomic spectrum are used to identify unknown atoms.

Robert Bunsen, a German chemist was the first to identify elements using line spectra. Using spectroscopic methods, we discovered elements such as helium, rubidium, scandium, gallium, thallium, etc.

Line Spectrum of Hydrogen

Hydrogen molecules dissociate when we pass electric discharge through gaseous hydrogen. Subsequently, the energetically excited H_2 atoms emit electromagnetic radiation of discrete frequencies giving rise to a spectrum. The hydrogen spectrum has many series of lines. These lines are named after their discoverers.

In 1885, the scientist Balmer showed that if spectral lines are expressed as wavenumber, then the visible lines of the hydrogen spectrum obey the following formula –

$$\text{Wavenumber} = 109{,}677\,(1/2^2 - 1/n^2)\ \text{cm}^{-1}$$

where $n \geq 3$.

We call this series of lines, Balmer series. These lines are the only lines in the hydrogen spectrum that appear in the visible region of electromagnetic radiation. Johannes Rydberg, a Swedish spectroscopist, showed that all series of lines in the hydrogen spectrum can be described by the formula –

$$\text{Wavenumber} = 109{,}677\,(1/n_1^2 - 1/n_2^2)\ \text{cm}^{-1}$$

where $n_1 = 1, 2....$ and $n_2 = n_1 + 1, n_1 + 2......$

The value 109,677 is the Rydberg constant for hydrogen. The lines that correspond to n_1 = 1, 2, 3, 4, 5 are called Lyman, Balmer, Paschen, Brackett and Pfund series, respectively.

Series	n_1	n_2	Spectral Region
Lyman	1	2,3...	Ultraviolet
Balmer	2	3,4...	Visible
Paschen	3	4,5...	Infrared
Brackett	4	5,6...	Infrared
Pfund	5	6,7...	Infrared

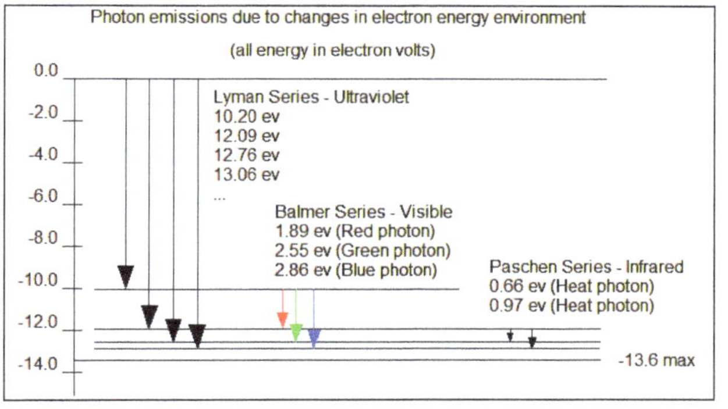

Photon emission due to changes in electron energy environment (all energy in electron volts).

The hydrogen atom has the simplest line spectrum of all elements. For heavier atoms, the line spectrum becomes more and more complex. However, there are certain features that are common to all line spectra –

- Line spectrum of every element is unique.
- There is regularity in the line spectrum of each element.

Now, that we understand the line spectrum of hydrogen, let's understand the features of the hydrogen atom, its structure, and its spectrum.

Bohr's Model for Hydrogen Atom

In 1913, Neils Bohr was the first to explain the features of the hydrogen atom and its spectrum. His model is based on the following postulates –

- Electrons in a hydrogen atom can move around the nucleus in a circular path that has a fixed radius and energy. These paths, arranged concentrically around the nucleus, are orbits.
- The energy of an electron in orbit does not change with time. However, when the electron absorbs a required amount of energy, it moves from a lower stationary state to a higher stationary state. On the other hand, when the electron emits energy, it moves from a higher stationary state to a lower stationary state. This energy change is not continuous.
- The following equation gives the frequency of radiation absorbed or emitted when an electron transitions between two stationary states that differ in energy by ΔE –

  $\nu = \Delta E/h = (E_2 - E_1)/h$

  Where, E_1 and E_2 are the energies of the lower and higher energy states, respectively. This equation is commonly called Bohr's frequency rule.

- In a given stationary state, the angular momentum of an electron is given as –

  $m_e v r = n.(h/2\pi)$ where n = 1,2,3......

  Thus, an electron can only move in those orbits whose angular momentum is an integral multiple of $h/2\pi$. This is why there are only certain orbits.

Hydrogen Atom described by Bohr's Model

For the hydrogen atom, Bohr's theory describes that –

- The stationary states for the electron are numbered n = 1,2,3... These are Principal Quantum numbers.
- The radii of the stationary states are given as $r_n = n^2 a_o$, where a_o = 52,9pm. Therefore, the radius of the first stationary state i.e. the Bohr orbit is 52.9pm. The electron in the hydrogen atom is in this orbit i.e. n=1. As *n* increases, *r* will increase and the electron will be found away from the nucleus.
- The energy of the stationary state is given by the equation –

  $E_n = - R_H (1/n^2)$ where n=1,2,3......

  R_H is the Rydberg constant and the value is 2.18×10^{-18} J. Using this equation, you can calculate the energy of any state. For example, the energy of the stationary state for n=2 will be,

  $E_2 = -2.18\times10^{-18} J (1/2^2) = -0.545\times10^{-18} J$

  The energy of an electron is taken as zero when it is not under the influence of the nucleus. In this situation, n=∞ and the atom are called an ionized hydrogen atom.

- Bohr's theory is also applicable to ions with only one electron, for example, the hydrogen atom. For these kinds of ions, the energies of stationary states are –

  $E_n = -2.18 \times 10^{-18} (Z^2/n^2) J$

  The radii are given by $r_n = 52.9(n^2)/Z$ pm, where Z is the atomic mass number.

- With Bohr's theory, it is also possible to calculate the velocities of electrons moving in orbits. The magnitude of the velocity of the electron increases with an increase in positive charge on the nucleus and decreases with an increase in principal quantum number.

Explanation of Line Spectrum of Hydrogen

Bohr's model can explain the line spectrum of the hydrogen atom. According to assumption 2, radiation is absorbed when an electron goes from orbit of lower energy to higher energy; whereas radiation is emitted when it moves from higher to lower orbit. The energy gap between the two orbits is –

$\Delta E = E_f - E_i$ where f is the final orbit, i is the initial orbit

Since, $E_n = -R_H (1/n^2)$, we can now say $\Delta E = (-R_H/n^2_f) - (-R_H/n^2_i) = R_H(1/n^2_i - 1/n^2_f)$

$= 2.18 \times 10^{-18} J (1/n^2_i - 1/n^2_f)$

The frequency and wavenumber associated with the absorption and emission of the photon can also be calculated –

$\nu = \Delta E/h = R_H(1/n^2_i - 1/n^2_f) = 2.18 \times 10^{-18} J/6.626 \times 10^{-34} J s (1/n^2_i - 1/n^2_f)$

$= 3.29 \times 10^{15} (1/n^2_i - 1/n^2_f)$ Hz

Wavenumber $= \nu/c = R_H/hc (1/n^2_i - 1/n^2_f) = 3.29 \times 10^{15} s^{-1}/ 3 \times 10^8 ms^{-s} (1/n^2_i - 1/n^2_f)$

$= 1.09677 \times 10^7 (1/n^2_i - 1/n^2_f) m^{-1}$

- When $n_f > n_i$, the term in brackets is positive and energy is absorbed.
- When $n_i > n_f$, ΔE is negative and energy is released.

Limitations of Bohr's Model

Bohr's model, although an improvement over Rutherford's model, failed to account for the following points –

- It fails to explain the finer details of the hydrogen atom spectrum observed by spectroscopic techniques. Bohr's model is unable to explain the spectrum of atoms other than hydrogen. It also fails to explain the concepts of Zeeman and the Stark effect.
- It does not explain the ability of atoms to form molecules with chemical bonds.

References

- Atom, science: britannica.com, Retrieved 10 May, 2019
- Atoms-to-Molecules-HS-ES, atoms-and-molecules, earth-science: ck12.org, Retrieved 12 August, 2019
- Relative-Atomic-Mass, GCSE, Chemistry: ivyroses.com, Retrieved 18 March, 2019
- Molecule: whatis.techtarget.com, Retrieved 27 June, 2019
- Molecular-mass-and-mole-concept, atoms-and-molecules, chemistry: toppr.com, Retrieved 23 April, 2019
- Atomic-structure: byjus.com, Retrieved 8 August, 2019

- Daltons-atomic-theory, some-basic-concepts-of-chemistry, chemistry: toppr.com, Retrieved 14 July, 2019
- Thomsons-model, chemistry: byjus.com, Retrieved 19 February, 2019
- Rutherfords-model-of-an-atom, structure-of-atom, chemistry: toppr.com, Retrieved June 25, 2019
- Bohrs-atomic-model-and-its-limitations, chemistry: byjus.com, Retrieved 16 March, 2019
- Bohr-model-of-the-atom: thoughtco.com, Retrieved 21 May, 2019
- Definition-of-periodic-law: thoughtco.com, Retrieved 10 June, 2019
- Pauli-exclusion-principle: chemicool.com, Retrieved 11 July, 2019
- Aufbau-principle, chemistry: byjus.com, Retrieved 24 January, 2019
- Hunds-rule, chemistry: byjus.com, Retrieved 13 August, 2019
- Emission-and-absorption-spectra, structure-of-atom, chemistry: toppr.com, Retrieved 9 April, 2019

3

Chemical Bonding and Reactivity

The attraction between atoms, ions and molecules that facilitates the formation of chemical compounds is known as chemical bond. The various types of chemical bonds include covalent bond, ionic bond, metallic bond, etc. A chemical reaction is a process that transforms one set of chemical substances to another. The topics elaborated in this chapter will help in gaining a better perspective of chemical bonding and reactivity.

Chemical Bond

Chemical bonding is any of the interactions that account for the association of atoms into molecules, ions, crystals, and other stable species that make up the familiar substances of the everyday world. When atoms approach one another, their nuclei and electrons interact and tend to distribute themselves in space in such a way that the total energy is lower than it would be in any alternative arrangement. If the total energy of a group of atoms is lower than the sum of the energies of the component atoms, they then bond together and the energy lowering is the bonding energy.

Types of Chemical Bonds

Ionic Bond

Ionic bonding involves a transfer of an electron, so one atom gains an electron while one atom loses an electron. One of the resulting ions carries a negative charge (anion), and the other ion carries a positive charge (cation). Because opposite charges attract, the atoms bond together to form a molecule.

Covalent Bond

The most common bond in organic molecules, a covalent bond involves the sharing of electrons between two atoms. The pair of shared electrons forms a new orbit that extends around the nuclei of both atoms, producing a molecule. There are two secondary types of covalent bonds that are relevant to biology — polar bonds and hydrogen bonds.

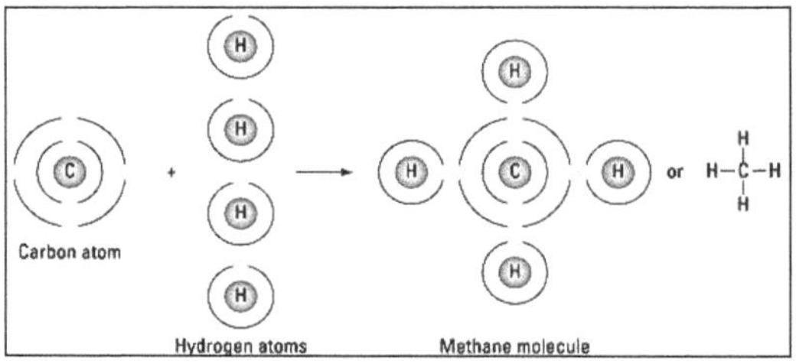

Polar Bond

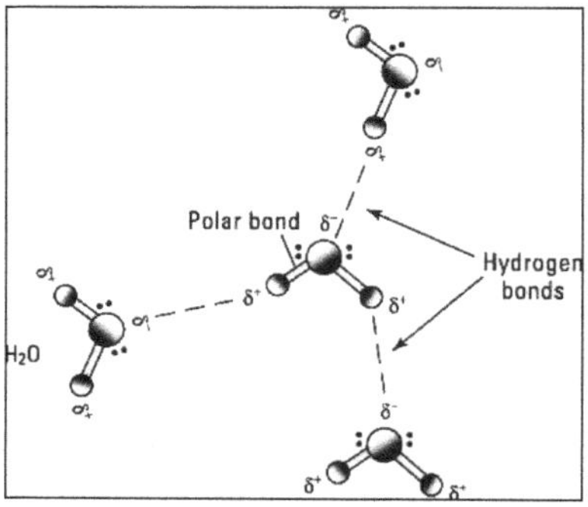

Two atoms connected by a covalent bond may exert different attractions for the electrons in the bond, producing an unevenly distributed charge. The result is known as a polar bond, an intermediate case between ionic and covalent bonding, with one end of the molecule slightly negatively charged and the other end slightly positively charged.

These slight imbalances in charge distribution are indicated in the figure by lowercase delta symbols with a charge superscript (+ or –). Although the resulting molecule is neutral, at close distances the uneven charge distribution can be important. Water is an example of a polar molecule; the oxygen end has a slight positive charge whereas the hydrogen ends are slightly negative. Polarity explains why some substances dissolve readily in water and others do not.

Hydrogen Bond

Because they're polarized, two adjacent H_2O (water) molecules can form a linkage known as a hydrogen bond, where the (electronegative) hydrogen atom of one H_2O molecule is electrostatically attracted to the (electropositive) oxygen atom of an adjacent water molecule.

Consequently, molecules of water join together transiently in a hydrogen-bonded lattice. Hydrogen bonds have only about 1/20 the strength of a covalent bond, yet even this force is sufficient to affect the structure of water, producing many of its unique properties, such as high surface tension, specific heat, and heat of vaporization. Hydrogen bonds are important in many life processes, such as in replication and defining the shape of DNA molecules.

Metallic Bond

Metallic bond is force that holds atoms together in a metallic substance. Such a solid consists of closely packed atoms. In most cases, the outermost electron shell of each of the metal atoms overlaps with a large number of neighbouring atoms. As a consequence, the valence electrons continually move from one atom to another and are not associated with any specific pair of atoms. In short, the valence electrons in metals, unlike those in covalently bonded substances, are non-localized, capable of wandering relatively freely throughout the entire crystal. The atoms that the electrons leave behind become positive ions, and the interaction between such ions and valence electrons gives rise to the cohesive or binding force that holds the metallic crystal together.

Many of the characteristic properties of metals are attributable to the non-localized or free-electron character of the valence electrons. This condition, for example, is responsible for the high electrical conductivity of metals. The valence electrons are always free to move when an electrical field is applied. The presence of the mobile valence electrons, as well as the non-directionality of the binding force between metal ions, account for the malleability and ductility of most metals. When a metal is shaped or drawn, it does not fracture, because the ions in its crystal structure are quite easily displaced with respect to one another. Moreover, the non-localized valence electrons act as a buffer between the ions of like charge and thereby prevent them from coming together and generating strong repulsive forces that can cause the crystal to fracture.

Valence Bond Theory

Heitler and London introduced this theory. This is primarily based on the concepts of atomic orbitals, electronic configuration of elements, the overlapping of atomic orbitals, hybridization of atomic orbitals. The overlapping of atomic orbitals results in the formation of a chemical bond. The electrons are localized in the bond region due to overlapping.

Valence bond theory describes the electronic structure of molecules. The theory says that electrons fill the atomic orbitals of an atom within a molecule. It also states that the nucleus of one atom is attracted to the electrons of another atom. Now, we move on and look at the various postulates of the valence bond theory.

Postulates of Valence Bond Theory

- The overlapping of two half-filled valence orbitals of two different atoms results in the formation of the covalent bond. The overlapping causes the electron density between two bonded atoms to increase. This gives the property of stability to the molecule.

- In case the atomic orbitals possess more than one unpaired electron, more than one bond can be formed and electrons paired in the valence shell cannot take part in such a bond formation.
- A covalent bond is directional. Such a bond is also parallel to the region of overlapping atomic orbitals.
- Based on the pattern of overlapping, there are two types of covalent bonds: sigma bond and a pi bond. The covalent bond formed by sidewise overlapping of atomic orbitals is known as pi bond whereas the bond formed by overlapping of atomic orbital along the inter nucleus axis is known as a sigma bond.

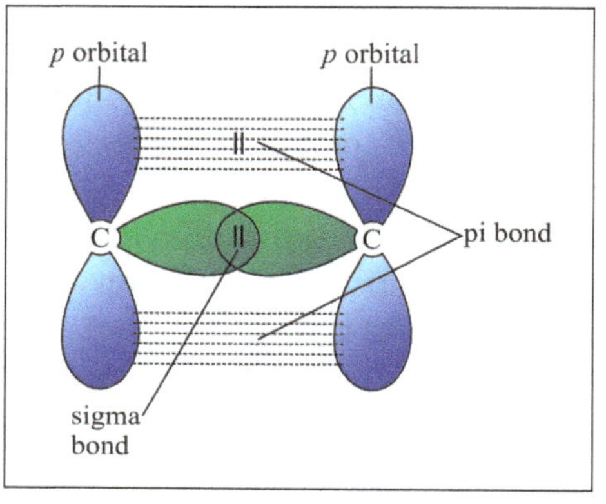

Limitations of Valence Bond Theory

As we pointed out earlier, nothing is perfect! In a similar way, the Valence Bond theory is also not perfect. It has its own set of limitations. They are:

- It fails to explain the tetravalency of carbon.
- This theory does not discuss the electrons' energies.
- The assumptions are about the electrons being localized to specific locations.

Molecular Orbital Theory

The Valence Bond Theory fails to answer certain questions like why He_2 molecule does not exist and why O_2 is paramagnetic. Therefore in 1932 F. Hood and R.S. Mulliken came up with Molecular Orbital Theory.

According to the Molecular Orbital Theory, individual atoms combine to form molecular orbitals. Thus the electrons of an atom are present in various atomic orbitals and are associated with several nuclei.

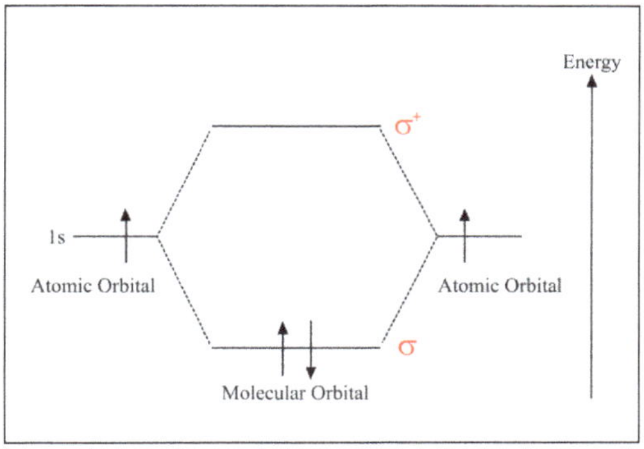

We know that we can consider electrons as either particle or wave nature. Therefore, we can describe an electron in an atom as occupying an atomic orbital, or by a wave function Ψ. These are solutions to the Schrodinger wave equation. Electrons in a molecule occupy molecular orbitals. We can obtain the wave function of a molecular orbital by the following methods.

- Linear Combination of Atomic Orbitals (LCAO)
- United Atom Method

Linear Combination of Atomic Orbitals (LCAO)

As per this method, the formation of orbitals is because of Linear Combination (addition or subtraction) of atomic orbitals which combine to form the molecule. Consider two atoms A and B which have atomic orbitals described by the wave functions Ψ_A and Ψ_B.

If the electron cloud of these two atoms overlaps, then we can obtain the wave function for the molecule by a linear combination of the atomic orbitals Ψ_A and Ψ_B. The below equation forms two molecular orbitals.

$$\Psi_{MO} = \Psi_A \pm \Psi_B$$

Bonding Molecular Orbitals

When the addition of wave function takes place, the type of molecular orbitals formed are Bonding Molecular Orbitals. We can represent them by $\Psi_{MO} = \Psi_A + \Psi_B$. They have lower energy than atomic orbitals involved.

Anti-Bonding Molecular Orbitals

When molecular orbital forms by the subtraction of wave function, the type of molecular orbitals formed are anti-bonding Molecular Orbitals. We can represent them as $\Psi_{MO} = \Psi_A + \Psi_B$. They have higher energy than atomic orbitals. Therefore, the combination of two atomic orbitals results in the formation of two molecular orbitals. They are the bonding molecular orbital (BMO) and the anti-bonding molecular orbital (ABMO).

Relative Energies of Molecular Orbitals

- Bonding Molecular Orbitals (BMO) – Energy of Bonding Molecular Orbitals is less than that of Anti Bonding Molecular Orbitals. This is because of the increase in the attraction of both the nuclei for both the electron (of the combining atom).
- Anti-Bonding Molecular Orbitals (ABMO) – Energy of Anti Bonding Molecular Orbitals is higher than Bonding Molecular Orbitals. This is because the electron tries to move away from the nuclei and are in a repulsive state.

Rules for Filling of Molecular Orbitals

We have to follow certain rules while filling up molecular orbitals with electrons in order to write correct molecular configurations. They are,

- Aufbau Principle – This principle states that those molecular orbitals which have the lowest energy are filled first.
- Pauli's Exclusion Principle – According to this principle, each molecular orbital can accommodate a maximum of two electrons having opposite spins.

VSEPR Theory

The Valence Shell Electron Pair Repulsion Model is often abbreviated as VSEPR (pronounced "vesper"). It is basically a model to predict the geometry of molecules. Specifically, VSEPR models look at the bonding and molecular geometry of organic molecules and polyatomic ions. It is useful for nearly all compounds that have a central atom that is not a metal.

Importance of VSEPR Models

- Lewis structures only tell the number and types of bonds between atoms, as they are limited to two dimensions. The VSEPR model predicts the 3-D shape of molecules and ions but is ineffective in providing any specific information regarding the bond length or the bond itself.
- VSEPR models are based on the concept that electrons around a central atom will configure themselves to minimize repulsion, and that dictates the geometry of the molecule.

- It can predict the shape of nearly all compounds that have a central atom, as long as the central atom is not a metal. Each shape has a name and an idealized bond angle associated with it.

The following terms are commonly used in discussing the shapes of molecules-

- Bond Angle: This is the angle between a bonded atom, the central atom, and another bonded atom.
- Lone Pair: This refers to a pair of valence electrons that are not shared with another atom.
- Molecular Geometry: This is the 3-D arrangement of bonded atoms in a polyatomic ion or molecule.
- Electron Pair Geometry: This is the 3-D arrangement of electron pairs around the central atom of a polyatomic ion or molecule.

The main difference between molecular geometry and electron pair geometry is that molecular geometry does not include unpaired electrons, whereas electron pair geometry includes both bonded atoms and unpaired electrons. If there are no unpaired electrons in the compound being assessed, the molecular and electron pair geometries will be the same.

Molecular Geometry

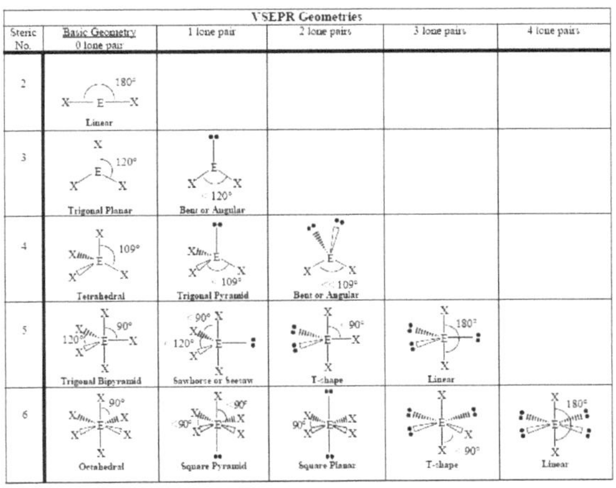

VSEPR Geometries

Steps to using VSEPR

- Draw a Lewis structure for the ion or molecule in question.
- Determine the number of electron groups around the central atom. Each lone pair of electrons counts as a single group. Each bond counts as a single group, even if it is a double or triple bond. Find the corresponding electron geometry from the table.

- Determine the number of lone pairs and the number of bonding pairs around the central atom, and use that to find the molecular geometry.

VSEPR Notation

VSEPR notation gives a general formula for classifying chemical species based on the number of electron pairs around a central atom. Note, however, that not all species have the same molecular geometry. For example, carbon dioxide and sulfur dioxide are both species, but one is linear and the other is bent.

Sometimes, the notation is expanded to include lone pair electrons. This can get confusing because water can be referred to as a species depending on the conventions the author or text chooses. In general,

- A is used to represent the central atom.
- B or X is used to represent the number of atoms bonded to the central atom.
- E represents the number of lone pairs on the central atom (ignore lone pairs on bonded atoms).

Again, this theory is also not void of any limitations. We will now discuss the common limitations of the VSEPR theory.

Limitations of the VSEPR Theory

- The VSEPR model is not a theory. It does not explain or attempt to explain any observations or predictions. Rather, it is an algorithm that accurately predicts the structures of a large number of compounds.
- VSEPR is simple and useful but does not work for all chemical species.
- First, the idealized bond angles do not always match the measured values. For example, VSEPR predicts that and will have the same bond angles, but structural studies have shown the bonds in the two molecules are different by 12 degrees.
- VSEPR also predicts that group-2 halides such as will be linear when they are actually bent. Quantum mechanics and atomic orbitals can give more sophisticated predictions when VSEPR is inadequate.

Chemical Reactivity

In chemistry, reactivity is a measure of how readily a substance undergoes a chemical reaction. The reaction can involve the substance on its own or with other atoms or compounds, generally accompanied by a release of energy. The most reactive elements and compounds may ignite spontaneously or explosively. They generally burn in water as well as the oxygen in the air. Reactivity is dependent upon temperature. Increasing temperature increases the energy available for a chemical reaction, usually making it more likely.

Another definition of reactivity is that it is the scientific study of chemical reactions and their kinetics.

Reactivity Trend in the Periodic Table

The organization of elements on the periodic table allows for predictions concerning reactivity. Both highly electropositive and highly electronegative elements have a strong tendency to react. These elements are located in the upper right and lower left corners of the periodic table and in certain element groups. The halogens, alkali metals, and alkaline earth metals are highly reactive.

- The most reactive element is fluorine, the first element in the halogen group.
- The most reactive metal is francium, the last alkali metal. However, francium is an unstable radioactive element, only found in trace amounts. The most reactive metal that has a stable isotope is cesium, which is located directly above francium on the periodic table.
- The least reactive elements are the noble gases. Within this group, helium is the least reactive element, forming no stable compounds.
- Metal can have multiple oxidation states and tend to have intermediate reactivity. Metals with low reactivity are called noble metals. The least reactive metal is platinum, followed by gold. Because of their low reactivity, these metals don't readily dissolve in strong acids. Aqua regia, a mixture of nitric acid and hydrochloric acid, is used to dissolve platinum and gold.

Working of Reactivity

A substance reacts when the products formed from a chemical reaction have lower energy (higher stability) than the reactants. The energy difference can be predicted using valence bond theory, atomic orbital theory, and molecular orbital theory. Basically, it boils down to the stability of electrons in their orbitals. Unpaired electrons with no electrons in comparable orbitals are the most likely to interact with orbitals from other atoms, forming chemical bonds. Unpaired electrons with degenerate orbitals that are half-filled are more stable but still reactive. The least reactive atoms are those with a filled set of orbitals (octet).

The stability of the electrons in atoms determines not only the reactivity of an atom but its valence and the type of chemical bonds it can form. For example, carbon usually has a valence of 4 and forms 4 bonds because its ground state valence electron configuration is half-filled at $2s^2$ $2p^2$. A simple explanation of reactivity is that it increases with the ease of accepting or donating an electron. In the case of carbon, an atom can either accept 4 electrons to fill its orbital or (less often) donate the four outer electrons. While the model is based on atomic behavior, the same principle applies to ions and compounds.

Reactivity is affected by the physical properties of a sample, its chemical purity, and the presence of other substances. In other words, reactivity depends on the context in which a substance is viewed. For example, baking soda and water are not particularly reactive, while baking soda and vinegar readily react to form carbon dioxide gas and sodium acetate.

Particle size affects reactivity. For example, a pile of corn starch is relatively inert. If one applies a direct flame to the starch, it's difficult to initiate a combustion reaction. However, if the corn starch is vaporized to make a cloud of particles, it readily ignites.

Sometimes the term reactivity is also used to describe how quickly a material will react or the rate of the chemical reaction. Under this definition the chance of reacting and the speed of the reaction are related to each other by the rate law:

$$\text{Rate} = k[A]$$

Where rate is the change in molar concentration per second in the rate-determining step of the reaction, k is the reaction constant (independent of concentration), and [A] is the product of the molar concentration of the reactants raised to the reaction order (which is one, in the basic equation). According to the equation, the higher the reactivity of the compound, the higher its value for k and rate.

Stability Versus Reactivity

Sometimes a species with low reactivity is called "stable", but care should be taken to make the context clear. Stability can also refer to slow radioactive decay or to the transition of electrons from the excited state to less energetic levels (as in luminescence). A nonreactive species may be called "inert". However, most inert species actually do react under the right conditions to form complexes and compounds (e.g., higher atomic number noble gases).

Chemical Reactions

A chemical reaction is a chemical change which forms new substances. A chemical reaction may be represented by a chemical equation, which indicates the number and type of each atom, as well as their organization into molecules or ions. A chemical equation uses the element symbols as shorthand notation for the elements, with arrows to indicate the direction of the reaction. A conventional reaction is written with reactants on the left side of the equation and products on the right side. The state of matter of the substances may be indicated in parenthesis (s for solid, l for liquid, g for gas, aq for aqueous solution). The reaction arrow may go from left to right or there may be a double arrow, indicating reactants turn to products and some product undergoes the reverse reaction to reform reactants.

While chemical reactions involves atoms, typically only the electrons are involved in the breaking and formation of chemical bonds. Processes involving the atomic nucleus are called nuclear reactions.

The substances that participate in a chemical reaction are called reactants. The substances that are formed are called products. The products have different properties from the reactants. Also Known as: reaction, chemical change.

Chemical Reaction Examples

The chemical reaction $H_2(g) + ½ O_2(g) \rightarrow H_2O(l)$ describes the formation of water from its elements.

The reaction between iron and sulfur to form iron(II) sulfide is another chemical reaction, represented by the chemical equation:

$$8\,Fe + S_8 \rightarrow 8\,FeS$$

Types of Chemical Reactions

Here are some common reaction types:

Oxidation-reduction or Redox Reaction

In a redox reaction, the oxidation numbers of atoms are changed. Redox reactions may involve the transfer of electrons between chemical species.

The reaction that occurs when I_n which I_2 is reduced to I^- and $S_2O_3^{2-}$ (thiosulfate anion) is oxidized to $S_4O_6^{2-}$ provides an example of a redox reaction:

$$2\,S_2O_3^{2-}(aq) + I_2(aq) \rightarrow S_4O6^{2-}(aq) + 2\,I^-(aq)$$

Direct Combination or Synthesis Reaction

In a synthesis reaction, two or more chemical species combine to form a more complex product.

$$A + B \rightarrow AB$$

The combination of iron and sulfur to form iron (II) sulfide is an example of a synthesis reaction:

$$8\ Fe + S_8 \rightarrow 8\ FeS$$

Chemical Decomposition or Analysis Reaction

In a decomposition reaction, a compound is broken into smaller chemical species.

$$AB \rightarrow A + B$$

The electrolysis of water into oxygen and hydrogen gas is an example of a decomposition reaction:

$$2\ H_2O \rightarrow 2\ H_2 + O_2$$

Single Displacement or Substitution Reaction

A substitution or single displacement reaction is characterized by one element being displaced from a compound by another element.

$$A + BC \rightarrow AC + B$$

An example of a substitution reaction occurs when zinc combines with hydrochloric acid. The zinc replaces the hydrogen:

$$Zn + 2\ HCl \rightarrow ZnCl_2 + H_2$$

Metathesis or Double Displacement Reaction

In a double displacement or metathesis reaction two compounds exchange bonds or ions in order to form different compounds.

$$AB + CD \rightarrow AD + CB$$

An example of a double displacement reaction occurs between sodium chloride and silver nitrate to form sodium nitrate and silver chloride.

$$NaCl(aq) + AgNO_3(aq) \rightarrow NaNO_3(aq) + AgCl(s)$$

Acid-base Reaction

An acid-base reaction is a type of double displacement reaction that occurs between an acid and a base. The H^+ ion in the acid reacts with the OH^- ion in the base to form water and an ionic salt:

$$HA + BOH \rightarrow H_2O + BA$$

The reaction between hydrobromic acid (HBr) and sodium hydroxide is an example of an acid-base reaction:

$$HBr + NaOH \rightarrow NaBr + H_2O$$

Combustion

A combustion reaction is a type of redox reaction in which a combustible material combines with an oxidizer to form oxidized products and generate heat (exothermic reaction). Usually, in a combustion reaction oxygen combines with another compound to form carbon dioxide and water. An example of a combustion reaction is the burning of naphthalene:

$$C_{10}H_8 + 12\ O_2 \rightarrow 10\ CO_2 + 4\ H_2O$$

Isomerization

In an isomerization reaction, the structural arrangement of a compound is changed but its net atomic composition remains the same.

Hydrolysis Reaction

A hydrolysis reaction involves water. The general form for a hydrolysis reaction is:

$$X^-(aq) + H_2O(l) \rightarrow HX(aq) + OH^-(aq).$$

References

- Chemical-bonding, science: britannica.com, Retrieved 8 February, 2019
- Chemical-bonding, science: britannica.com, Retrieved 8 February, 2019
- 4-types-of-chemical-bonds, anatomy, science: dummies.com, Retrieved 15 April, 2019
- Metallic-bond, science: britannica.com, Retrieved 17 June, 2019
- Valence-bond-theory, chemical-bonding-and-molecular-structure, chemistry: toppr.com, Retrieved 23 January, 2019
- Molecular-orbital-theory, chemical-bonding-and-molecular-structure, chemistry: toppr.com, Retrieved 12 July, 2019
- Vsepr-theory, chemical-bonding-and-molecular-structure, chemistry: toppr.com, Retrieved 10 May, 2019
- Reactivity-definition: thoughtco.com, Retrieved 20 August, 2019
- Chemical-reaction-definition: thoughtco.com, Retrieved 13 June, 2019
- Types-of-chemical-reactions: thoughtco.com, Retrieved 19 March, 2019

4

Chemical Kinetics

Chemical kinetics is the domain of physical chemistry that studies the rates of chemical reactions. It focuses on rate laws and rate constants, reaction mechanism, collision theory, activation energy, etc. The chapter closely examines these key concepts of chemical kinetics to provide an extensive understanding of the subject.

Chemical kinetics is the branch of physical chemistry that is concerned with understanding the rates of chemical reactions. It is to be contrasted with thermodynamics, which deals with the direction in which a process occurs but in itself tells nothing about its rate. Thermodynamics is time's arrow, while chemical kinetics is time's clock. Chemical kinetics relates to many aspects of cosmology, geology, biology, engineering, and even psychology and thus has far-reaching implications. The principles of chemical kinetics apply to purely physical processes as well as to chemical reactions.

One reason for the importance of kinetics is that it provides evidence for the mechanisms of chemical processes. Besides being of intrinsic scientific interest, knowledge of reaction mechanisms is of practical use in deciding what is the most effective way of causing a reaction to occur. Many commercial processes can take place by alternative reaction paths, and knowledge of the mechanisms makes it possible to choose reaction conditions that favour one path over others.

A chemical reaction is, by definition, one in which chemical substances are transformed into other substances, which means that chemical bonds are broken and formed so that there are changes in the relative positions of atoms in molecules. At the same time, there are shifts in the arrangements of the electrons that form the chemical bonds. A description of a reaction mechanism must therefore deal with the movements and speeds of atoms and electrons. The detailed mechanism by which a chemical process occurs is referred to as the reaction path, or pathway.

The vast amount of work done in chemical kinetics has led to the conclusion that some chemical reactions go in a single step; these are known as elementary reactions. Other reactions go in more than one step and are said to be stepwise, composite, or complex. Measurements of the rates of chemical reactions over a range of conditions can show whether a reaction proceeds by one or more steps. If a reaction is stepwise, kinetic measurements provide evidence for the mechanism of the individual elementary steps. Information about reaction mechanisms is also provided by certain non-kinetic studies, but little can be known about a mechanism until its kinetics has been investigated.

Even then, some doubt must always remain about a reaction mechanism. An investigation, kinetic or otherwise, can disprove a mechanism but can never establish it with absolute certainty.

Reaction Rate

The rate of a reaction is defined in terms of the rates with which the products are formed and the reactants (the reacting substances) are consumed. For chemical systems it is usual to deal with the concentrations of substances, which is defined as the amount of substance per unit volume. The rate can then be defined as the concentration of a substance that is consumed or produced in unit time. Sometimes it is more convenient to express rates as numbers of molecules formed or consumed in unit time.

The half-life

A useful rate measure is the half-life of a reactant, which is defined as the time that it takes for half of the initial amount to undergo reaction. For a special type of kinetic behaviour, the half-life is independent of the initial amount. A common and straightforward example of a half-life independent of the initial amount is radioactive substances. Uranium-238, for example, decays with a half-life of 4.5 billion years; of an initial amount of uranium, half of that amount will have decayed in that period of time. The same behaviour is found in many chemical reactions.

Even when the half-life of a reaction varies with the initial conditions, it is often convenient to quote a half-life, bearing in mind that it applies only to the particular initial conditions. Consider, for example, the reaction in which hydrogen and oxygen gases combine to form water; the chemical equation is

$$2H_2 + O_2 \rightarrow 2H_2O.$$

If the gases are mixed together at atmospheric pressure and room temperature, nothing observable will happen over long periods of time. However, reaction does occur, with a half-life that is estimated to be more than 12 billion years, which is roughly the age of the universe. If a spark is passed through the system, the reaction occurs with explosive violence, with a half-life of less than one-millionth of a second. This is a striking example of the great range of rates with which chemical kinetics is concerned. There are many possible processes that proceed too slowly to be studied experimentally, but sometimes they can be accelerated, often by the addition of a substance known as a catalyst. Some reactions are even faster than the hydrogen-oxygen explosion—for example, the combination of atoms or molecular fragments (called free radicals) where all that occurs is the formation of a chemical bond. Some modern kinetic investigations are concerned with even faster processes, such as the breakdown of highly energetic and therefore transient molecules, where times of the order of femtoseconds (fs; 1 fs = 10^{-15} second) are involved.

Measuring Slow Reactions

The best way to study exceedingly slow reactions is to change the conditions so that the reactions occur in a reasonable time. Increasing the temperature, which can have a strong effect on the reaction rate, is one possibility. If the temperature of a hydrogen-oxygen mixture is raised to about 500 °C (900 °F), reaction then occurs rapidly, and its kinetics has been studied under those conditions. When a reaction occurs to a measurable extent over a period of minutes, hours, or days, rate measurements are straightforward. Amounts of reactants or products are measured at various times, and the rates are readily calculated from the results. Many automated systems have now been devised for measuring rates in this way.

Measuring Fast Reactions

Some processes are so fast that special techniques have to be used to study them. There are two difficulties with fast reactions. One is that the time that it takes to mix reactants or to change the temperature of the system may be significant in comparison with the half-life, so that the initial time cannot be measured accurately. The other difficulty is that the time it takes to measure the amounts of substances may be comparable with the half-life of the reaction. The methods used to overcome these difficulties fall into two classes: flow methods and pulse and probe methods.

In flow methods, two gases or solutions are introduced rapidly into a mixing vessel, and the resulting mixture then flows rapidly along a tube. Concentrations of reactants or products may then be measured—for example, by spectroscopic methods—at various positions along the tube, which correspond to various reaction times. A modification of this method is the stopped-flow technique, in which the reactants are forced rapidly into a reaction chamber; the flow is then suddenly stopped, and the amounts are measured by physical methods after various short times. These flow methods are limited by the time it takes to mix gases or solutions and are not suitable if the half-life is less than about a hundredth of a second.

These mixing difficulties were overcome by pulse and probe methods. The principle of these is that a short pulse, usually of radiation, is given to a chemical system and is then followed by a probe, usually involving radiation that provides spectroscopic evidence of what occurred after the initial pulse. The first of these methods, developed in 1949 by British chemists R.G.W. Norrish and George Porter, was the flash-photolysis method, for which Norrish and Porter won the Nobel Prize for Chemistry in 1967. In this technique a flash of light of high intensity but short duration brings about the formation of atomic and molecular species, the reactions of which can be studied kinetically by spectroscopy. In the earliest experiments the duration of the flash was about a millisecond (ms; 1 ms = 10^{-3} second), but in the next four decades the duration was reduced by more than 11 powers of 10, to just a few femtoseconds. A nanosecond (ns; 1 ns = 10^{-9} second) flash is adequate for studying almost any purely chemical reaction where there is a change in chemical identity. Any chemical reaction, however, involves processes of a purely physical nature, such as energy redistribution and the breakdown of transient species, which occur in the femtosecond range.

Many such processes have now been studied with flashes of only a few femtoseconds' duration. The time that it takes for the length of a chemical bond to change by 10^{-10} metre can be as little as about 100 fs, so that a flash of a few femtoseconds' duration, closely followed by another one of the same duration, will provide information about such tiny changes in bond lengths. The technique for causing one flash to occur a few nanoseconds after another is to route the light by a slightly longer path. A path of 1 additional micrometre (μm; 1 μm = 10^{-6}metre) causes a delay of 1 fs, and such a short path difference is now technically feasible. Egyptian-born chemist Ahmed Zewail won the Nobel Prize for Chemistry in 1999 for his work in this field.

Another pulse method is the relaxation method, developed in the 1950s by German physicist Manfred Eigen In this method the investigation begins with a reaction system in equilibrium; the reaction to be studied has finished, and no further changes take place. The external conditions are then altered very rapidly; the system is then no longer at equilibrium, and it relaxes to a new equilibrium. The speed of relaxation is measured by a physical method such as spectroscopy, and analysis of the results leads to the reaction rate.

The most common way of changing the external conditions is to change the temperature, and the method is called the temperature-jump, or T-jump, method. Techniques have been developed for raising the temperature of a tiny reaction vessel by a few degrees in less than 100 ns. The method is therefore not suitable for the fastest processes, which can be studied by flash photolysis, but many purely chemical processes are suitable for the T-jump technique, which has provided valuable kinetic information.

Other experimental techniques are used for the study of rapid processes. Ultrasonic methods have been used for processes occurring with half-lives in the microsecond (μs; 1 μs = 10^{-6}second) and nanosecond ranges. Nuclear magnetic resonance has also been used for certain types of reactions.

Some Kinetic Principles

The kinetic behaviour of an ordinary chemical reaction is conventionally studied in the first instance by determining how the reaction rate is influenced by certain external factors such as the concentrations of the reacting substances, the temperature, and sometimes the pressure. For a reaction in which two substances A and B react with each other, it is sometimes found that the reaction rate is proportional to the concentration of A, represented by [A], and to the concentration of B, or [B]. In that case the reaction is said to be a second-order reaction; it is first order in [A] and first order in [B]. In such a case the reaction rate v can be expressed as,

$$v = k[\mathrm{A}][\mathrm{B}],$$

Where k is a constant, and known as the rate constant for the reaction.

This is just one of many types of kinetics that can be observed. A substance A that changes into another substance may obey a kinetic equation of the form $v = k[\mathrm{A}]$, which is a first-order reaction. It is important to recognize that the kinetics of a reaction does not always correspond in a simple way to the balanced chemical equation for the reaction.

$$\mathrm{A} + \mathrm{B} \rightleftharpoons \mathrm{Y} + \mathrm{Z},$$

Thus, if a reaction is of the form, then it's not necessarily second-order in both directions. This is in contrast to the situation with the equilibrium constant for the reaction, which corresponds to the balanced equation. The reason why the kinetic law is different is that the reactions in the forward and reverse directions may occur by stepwise mechanisms that lead to a different and usually more complex kinetic equation.

Sometimes reaction rates depend on reactant concentrations in a more complicated way. This is a clear indication that a reaction happens in several steps.

The effect of temperature on reaction rates provides much information about reaction mechanisms. Understanding of this effect owes much to the ideas of the Dutch physical chemist Jacobus Henricus van't Hoff and the Swedish chemist Svante August Arrhenius.

Their equation for the dependence of a rate constant k on the absolute temperature T is,

$$k = A \exp(-E / RT),$$

Where R is the molar gas constant and A and E are quantities that are different for each reaction. This equation has come to be called the Arrhenius equation, although, as Arrhenius acknowledged when he applied it in 1889, it was first suggested by van't Hoff in 1884. According to this relationship, a plot of the logarithm of the rate constant against the reciprocal of the absolute temperature should yield a straight line. From the slope and intercepts of the line, it is possible to calculate the value of the kinetic parameters A and E. The Arrhenius relationship applies satisfactorily to most reactions and indeed to many physical processes; however, various complications may cause it to fail.

If the reaction between two molecules is an elementary one, occurring in a single step, a simple interpretation of the Arrhenius equation can be given. The quantity A is related to the frequency of collisions between the reacting molecules. The quantity E, known as the activation energy for the reaction, results from the fact that there is an energy barrier to reaction. If E was zero, k would be equal to A, which means that the reaction would occur every time a collision occurred between the reactant molecules. This is the case for reactions in which no chemical bond is broken, such as the combination of atoms.

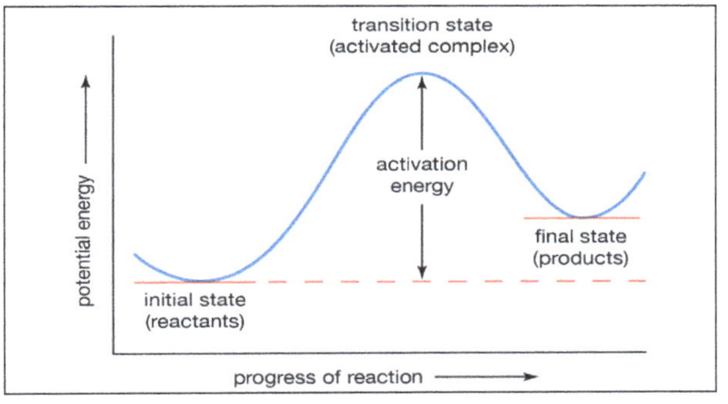

In the Arrhenius equation, the activation energy (E) represents the minimum amount of energy required to transform reactants into products in a chemical reaction. On a potential energy curve, the value of the activation energy is equivalent to the difference in potential energy between particles in an intermediate configuration (known as the activated complex, or transition state) and particles of reactants in their initial state. The activation energy thus can be visualized as a barrier that must be overcome by reactants before products can be formed.

For reactions in which a chemical bond is broken, on the other hand, the activation energy E is not zero but has a value that is often a tenth or so of the energy required to break the bond. A simple and essentially correct explanation of the activation energy was suggested by Arrhenius, who pointed out that, for many reactions, raising the temperature by 10 °C (18 °F) doubles the reaction rate. This increase cannot be caused by the increase in the frequency of collisions between colliding molecules, since the frequency does not increase sufficiently with a rise in temperature. Arrhenius suggested that when reactants A and B react together, they first form a highly energized intermediate that is denoted as $AB^{\ddagger}$, which subsequently gives the products of reaction:

$$A + B \rightleftharpoons AB^{\ddagger} \rightarrow Y + Z.$$

If the intermediate complex (also called the activated complex) $AB^{\ddagger}$ is of high energy, it is formed only in small amounts. According to the Boltzmann principle, the fraction of molecules having energy greater than E is exp $(-E/RT)$, which provides an explanation of the appearance of this fraction in the Arrhenius equation. The interpretation of the equation is thus that only those molecules having energy greater than E are able to undergo reaction; other collisions are ineffective, and the reactant molecules merely separate unchanged.

Rate Laws and Rate Constants

Experimental data is used to find reaction rates, from which rate laws and chemical kinetics rate constants are derived by applying the law of mass action. Rate laws allow for simple calculations for zero order reactions, first order reactions, and second order reactions.

- The rate of a zero-order reaction is constant and independent of the concentration of reactants. rate = k;
- The rate of a first-order reaction is proportional to the concentration of one reactants:rate = k[A];
- Therateofasecondorderreactionhasarateproportionaltothesquareoftheconcentration of a single reactant or else the product of the concentration of two reactants. rate = $k[A]^2$ or k[A][B].

Rate laws for individual steps must be combined to derive laws for more complex chemical reactions. For these reactions:

- There is a rate-determining step that limits the kinetics.
- The Arrhenius equation and Eyring equations may be used to experimentally determine activation energy.
- Steady-state approximations may be applied to simplify the rate law.

Factors that Affect Chemical Reaction Rate

Chemical kinetics predicts the rate of a chemical reaction will be increased by factors that increase the kinetic energy of the reactants (up to a point), leading to increased likelihood the reactants will interact with each other. Similarly, factors that decrease the chance of reactants colliding with each other may be expected to lower the reaction rate. The main factors that affect reaction rate are:

- Concentration of reactants (increasing concentration increases reaction rate);
- Temperature (increasing temperature increases reaction rate, up to a point);
- Presence of catalysts (catalysts offer a reaction a mechanism that requires a lower activation energy, so the presence of a catalyst increases the rate of a reaction);
- Physical state of reactants (reactants in the same phase may come into contact via thermal action, but surface area and agitation affect reactions between reactants in different phases);

- Pressure (for reactions involving gases, raising pressure increases the collisions between reactants, increasing reaction rate).

While chemical kinetics can predict the rate of a chemical reaction, it does not determine the extent to which the reaction occurs. Thermodynamics is used to predict equilibrium.

Factors Affecting Reaction Rate

Nature of the Reactants

The reaction rate varies depending upon what substances are reacting. Acid/base reactions, the formation of salts, and ion exchange are usually fast reactions. When covalent bond formation takes place between the molecules and when large molecules are formed, the reactions tend to be slower.

The nature and strength of bonds in reactant molecules greatly influence the rate of their transformation into products.

Physical State

The physical state (solid, liquid, or gas) of a reactant is also an important factor of the rate of change. When reactants are in the same phase, as in aqueous solution, thermal motion brings them into contact. However, when they are in different phases, the reaction is limited to the interface between the reactants. Reaction can occur only at their area of contact; in the case of a liquid and a gas, at the surface of the liquid. Vigorous shaking and stirring may be needed to bring the reaction to completion. This means that the more finely divided a solid or liquid reactant the greater its surface area per unit volume and the more contact it with the other reactant, thus the faster the reaction. To make an analogy, for example, when one starts a fire, one uses wood chips and small branches — one does not start with large logs right away. In organic chemistry, on water reactions are the exception to the rule that homogeneous reactions take place faster than heterogeneous reactions.

Surface Area of Solids

In a solid, only those particles that are at the surface can be involved in a reaction. Crushing a solid into smaller parts means that more particles are present at the surface, and the frequency of collisions between these and reactant particles increases, and so reaction occurs more rapidly. For example, Sherbet (powder) is a mixture of very fine powder of malic acid (a weak organic acid) and sodium hydrogen carbonate. On contact with the saliva in the mouth, these chemicals quickly dissolve and react, releasing carbon dioxide and providing for the fizzy sensation. Also, fireworks manufacturers modify the surface area of solid reactants to control the rate at which the fuels in fireworks are oxidised, using this to create different effects. For example, finely divided aluminium confined in a shell explodes violently. If larger pieces of aluminium are used, the reaction is slower and sparks are seen as pieces of burning metal are ejected.

Concentration

The reactions are due to collisions of reactant species. The frequency with which the molecules or ions collide depends upon their concentrations. The more crowded the molecules are, the more likely they are to collide and react with one another. Thus, an increase in the concentrations of the reactants will usually result in the corresponding increase in the reaction rate, while a decrease in the concentrations will usually have a reverse effect. For example, combustion will occur more rapidly in pure oxygen than in air (21% oxygen).

The rate equation shows the detailed dependence of the reaction rate on the concentrations of reactants and other species present. Different mathematical forms are possible depending on the reaction mechanism. The actual rate equation for a given reaction is determined experimentally and provides information about the reaction mechanism. The mathematical expression of the rate equation is often given by,

$$v = \frac{\mathrm{d}c}{\mathrm{d}t} = k\prod_i c_i^{m_i}$$

Here k is the reaction rate constant, c_i is the molar concentration of reactant i and m_i is the partial order of reaction for this reactant. The partial order for a reactant can only be determined experimentally and is often different from its stoichiometric coefficient.

Temperature

Temperature usually has a major effect on the rate of a chemical reaction. Molecules at a higher temperature have more thermal energy. Although collision frequency is greater at higher temperatures, this alone contributes only a very small proportion to the increase in rate of reaction. Much more important is the fact that the proportion of reactant molecules with sufficient energy to react (energy greater than activation energy: $E > E_a$) is significantly higher and is explained in detail by the Maxwell–Boltzmann distribution of molecular energies.

The 'rule of thumb' that the rate of chemical reactions doubles for every 10 °C temperature rise is a common misconception. This may have been generalized from the special case of biological systems, where the α (temperature coefficient) is often between 1.5 and 2.5.

A reaction's kinetics can also be studied with a temperature jump approach. This involves using a sharp rise in temperature and observing the relaxation time of the return to equilibrium. A particularly useful form of temperature jump apparatus is a shock tube, which can rapidly jump a gas's temperature by more than 1000 degrees.

Catalysts

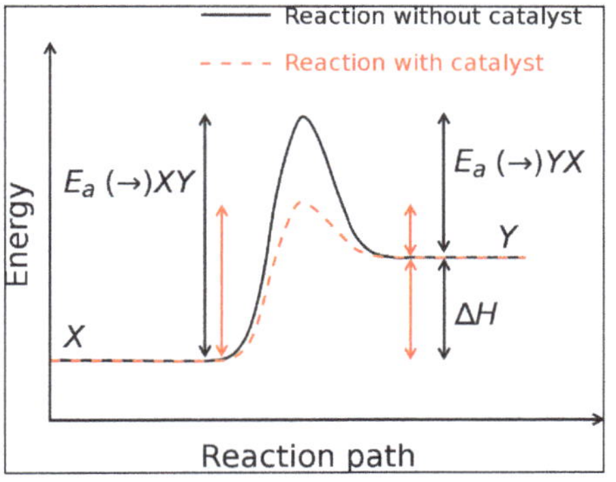

Generic potential energy diagram showing the effect of a catalyst in a hypothetical endothermic chemical reaction. The presence of the catalyst opens a different reaction pathway (shown in red) with a lower activation energy. The final result and the overall thermodynamics are the same.

A catalyst is a substance that alters the rate of a chemical reaction but remains chemically unchanged afterwards. The catalyst increases the rate of the reaction by providing a different reaction mechanism to occur with a lower activation energy. In autocatalysis a reaction product is itself a catalyst for that reaction leading to positive feedback. Proteins that act as catalysts in biochemical reactions are called enzymes. Michaelis–Menten kinetics describe the rate of enzyme mediated reactions. A catalyst does not affect the position of the equilibrium, as the catalyst speeds up the backward and forward reactions equally.

In certain organic molecules, specific substituents can have an influence on reaction rate in neighbouring group participation.

Pressure

Increasing the pressure in a gaseous reaction will increase the number of collisions between reactants, increasing the rate of reaction. This is because the activity of a gas is directly proportional to the partial pressure of the gas. This is similar to the effect of increasing the concentration of a solution.

In addition to this straightforward mass-action effect, the rate coefficients themselves can change due to pressure. The rate coefficients and products of many high-temperature gas-phase reactions change if an inert gas is added to the mixture; variations on this effect are called fall-off and chemical activation. These phenomena are due to exothermic or endothermic reactions occurring faster than heat transfer, causing the reacting molecules to have non-thermal energy distributions (non-Boltzmann distribution). Increasing the pressure increases the heat transfer rate between the reacting molecules and the rest of the system, reducing this effect.

Condensed-phase rate coefficients can also be affected by (very high) pressure; this is a completely different effect than fall-off or chemical-activation. It is often studied using diamond anvils.

A reaction's kinetics can also be studied with a pressure jump approach. This involves making fast changes in pressure and observing the relaxation time of the return to equilibrium.

Absorption of Light

The activation energy for a chemical reaction can be provided when one reactant molecule absorbs light of suitable wavelength and is promoted to an excited state. The study of reactions initiated by light is photochemistry, one prominent example being photosynthesis.

Reaction Mechanisms

The mechanism of a chemical reaction is the sequence of events that take place as reactant molecules are converted into products. The study of kinetics includes very complex and sophisticated reactions that cannot be analyzed without a proposed mechanism, a series of steps that a reaction takes before reaching the final products.

Reaction mechanisms are step-by-step descriptions of what occurs on a molecular level in chemical reactions. Each step of the reaction mechanism is known as an elementary process, a term used to describe a moment in the reaction when one or more molecules changes geometry or is perturbed by the addition or omission of another interacting molecule. Collectively, an overall reaction and a reaction mechanism consist of multiple elementary processes. These elementary steps are the basic building blocks of a complex reaction, and cannot be broken down any further.

A reaction mechanism is only a guess at how a reaction proceeds. Therefore, even if a mechanism agrees with the experimental results of a reaction, it cannot be proven to be correct.

Example

Consider the chlorination reaction of methane, CH_4:

$$CH_4(g) + 2Cl_2(g) \rightarrow CH_3Cl(g) + HCl(g) + Cl^-(g)$$

This reaction is proposed to occur via two successive elementary steps. Each step has its own characteristic reactants, product, rate law.

$$CH_4(g) + Cl_2(g) \rightarrow CH_3(g) + HCl(g)$$

with an elementary rate law of,

$$k_1 = [CH_4][Cl_2]$$

$$CH_3(g) + Cl_2(g) \rightarrow CH_3Cl(g) + Cl^-(g)$$

with an elementary rate law of,

$$k_2 = [CH_4][Cl_2]$$

The steps combine to generate the final reaction equation,

$$CH_4(g) + 2Cl_2(g) \rightarrow CH_3Cl(g) + HCl(g) + Cl^-(g)$$

Example

Consider the following reaction:

$$CO(g) + NO_2(g) \rightarrow CO_2(g) + NO(g)$$

This reaction equation suggests that the carbon monoxide directly reacts with nitrogen dioxide to form the products. However, its reaction mechanism shows this is not the case. Consider the two elementary processes below:

$$2NO_2(g) \rightarrow NO_3(g) + NO(g)$$
$$CO(g) + NO_3(g) \rightarrow CO_2(g) + NO_2(g)$$
$$CO(g) + NO_2(g) \rightarrow CO_2(g) + NO(g)$$

This reaction mechanism indicates that the compound NO_3 performs a necessary role in allowing the reaction to take place, a concept not possible to confirm by looking at the overall equation.

Example

Take the reaction $2NO(g) + O_2(g) \rightarrow 2NO_2(g)$.

A look at the overall reaction might suggest that two NO molecules must collide with an oxygen molecule to generate the product. However, termolecular processes are extremely rare; a further analysis of the reaction mechanism is required. The two elementary reactions below show:

$$2NO(g) \rightarrow N_2O_2(g)$$
$$2N_2O_2(g) + O_2(g) \rightarrow 2NO_2(g)$$
$$2NO(g) + O_2(g) \rightarrow 2NO_2(g)$$

The reaction mechanism shows that, in fact, three molecules combining simultaneously to form a product does not occur in this case. Instead, the overall reaction is a compilation of two individual elementary processes that take place.

Description of a Reaction Mechanism

Because a reaction mechanism is used to describe what occurs at each step of a reaction, it also describes the transition state, or the state in which the maximum of potential energy is reached. A mechanism must show the order in which the bonds form or break and the rate of each elementary step. Also accounted for are the reaction intermediates, stable molecules that do not appear in the experimentally determined rate law because they are formed in one step and consumed in a subsequent step. Because a reaction cannot proceed faster than the rate of slowest elementary step, the slowest step in a mechanism establishes the rate of the overall reaction. This elementary step is known as the rate-determining step.

A mechanism must satisfy the following two requirements:

1. The elementary steps must add up to give the overall balanced equation for the reaction.
2. The rate law for the rate-determining step must agree with the experimentally determined rate law.

Each of these events constitutes an elementary step that can be represented as a coming-together of discrete particles ("collision") or as the breaking-up of a molecule ("dissociation") into simpler units. The molecular entity that emerges from each step may be a final product of the reaction, or it might be an intermediate.

Collision Theory

Collision theory basically explains how reactions occur and why different reactions have different reactions rates. It states that,

- Molecules must collide in order to react.
- In order to effectively initiate a reaction, the molecules in the collisions must have sufficient energy to bring about disruptions in the bonds of molecules.
- A rise in temperature will cause the molecules to move faster and collide more vigorously, increasing the likelihood of bond cleavages and rearrangements greatly.
- The reactions containing neutral molecules cannot take place at all until they have acquired the activation energy needed to stretch, bend or distort one or more bonds.

Activation Energy

Activation energy is the energy that must be overcome in order for a reaction to occur. It is the minimum energy that is required to start a chemical reaction.

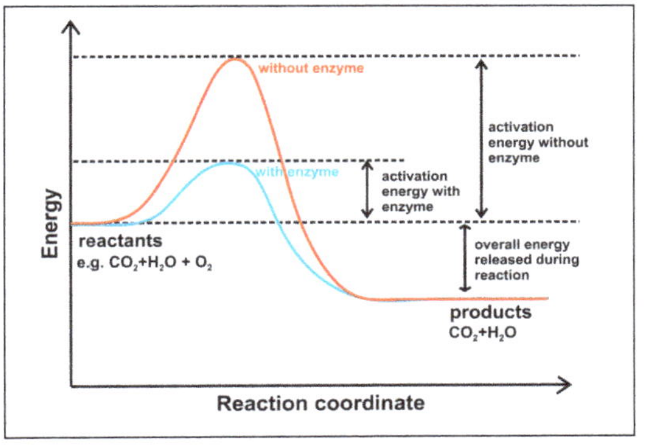

Explanation of Collision Theory

Collision theory qualitatively explains how chemical reactions occur and why different reactions have different rates. Consider a simple biomolecular step:

- Clearly, if two molecules A and B are to react, they must approach closely enough to disrupt some of their existing bonds and to permit the creation of new bonds that are required to form products. We call this a collision. The frequency of collision between A and B in a gas will be proportionate to the concentration of each. If we double the concentration of A, the frequency of A-B collision will double. Doubling the concentration of B will have the same effect.

- It is not enough that the molecules just collide. They need to be oriented in a specific manner that is appropriate for the process to occur. The molecules must collide with one another from the correct side. If they do not do so, the collision will not lead to the reaction.

- The molecules must collide with energies greater than or equal to the activation energy of the reaction. If this does not happen the reaction will not take place. The molecules need the energy to break their existing bonds and form new bonds. This is the kinetic energy that the molecules possess. If this energy is not equal to or greater than the activation energy, the reaction will not proceed.

Temperature Dependence of Collison theory

Thermal energy relates direction to motion at the molecular level. As the temperature rises, the molecules move faster and collide more vigorously, therefore causing more collisions and increasing the likelihood of bond cleavages. In most cases, the activation energy is supplied in the form of thermal energy.

As the reaction is completing and products are being formed, the activation energy is returned in the form of vibrational energy which is quickly released as heat. Therefore, it very important for the molecules to collide with energies greater than or equal to the activation energy of the reaction.

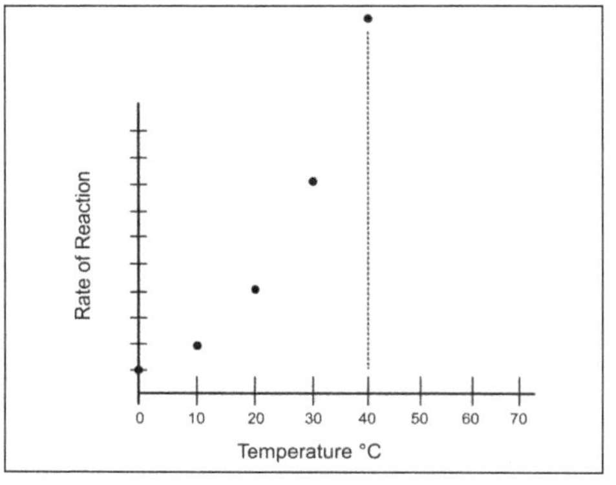

Rate of Reaction According to Collison Theory

For a bimolecular elementary reaction, A + B → Products, the rate of reaction is,

$$Rate = k\left[C_{12}H_{22}O_{11}\right]$$

Where Z_{AB} represents the collision frequency of reactants A and B and $^{-E_a/RT}$ represents the fraction of molecules with energies equal to or greater than the activation energy of the reaction. This is why different reactions have different reactions rates. Different reactions have different frequencies of reactants and different activation energies.

A collision that satisfies all the conditions in the collision theory and succeeds in forming a new product is known as an effective collision. Thus, the two important criteria in collision theory are the activation energy and proper orientation of molecules.

Activation Energy

Activation energy is the minimum amount of energy required to initiate a reaction. It is the height of the potential energy barrier between the potential energy minima of the reactants and products. Activation energy is denoted by E_a and typically has units of kilojoules per mole (kJ/mol) or kilocalories per mole (kcal/mol). The term "activation energy" was introduced by the Swedish scientist Svante Arrhenius in 1889. The Arrhenius equation relates activation energy to the rate at which a chemical reaction proceeds:

$$k = Ae^{-Ea/(RT)}$$

where k is the reaction rate coefficient, A is the frequency factor for the reaction, e is the irrational number (approximately equal to 2.718), E_a is the activation energy, R is the universal gas constant, and T is the absolute temperature (Kelvin).

From the Arrhenius equation, it can be seen that the rate of reaction changes according to temperature. Normally, this means a chemical reaction proceeds more quickly at a higher temperature. There are, however, a few cases of "negative activation energy", where the rate of a reaction decreases with temperature.

Need of Activation Energy

If you mix together two chemicals, only a small number of collisions will naturally occur between the reactant molecules to make products. This is particularly true if the molecules have low kinetic energy. So, before a significant fraction of reactants can be converted into products, the free energy of the system must be overcome. The activation energy gives the reaction that little extra push needed to get going. Even exothermic reactions require activation energy to get started. For example, a stack of wood won't start burning on its own. A lit match can provide the activation energy to start combustion. Once the chemical reaction starts, the heat released by the reaction provides the activation energy to convert more reactant into product.

Sometimes a chemical reaction proceeds without adding any additional energy. In this case, the activation energy of the reaction is usually supplied by heat from the ambient temperature. Heat increases the motion of the reactant molecules, improving their odds of colliding with each other and increasing the force of the collisions. The combination makes it more likely bonds between reactant will break, allowing for the formation of products.

Catalysts and Activation Energy

A substance that lowers the activation energy of a chemical reaction is called a catalyst. Basically, a catalyst acts by modifying the transition state of a reaction. Catalysts are not consumed by the chemical reaction and they don't change the equilibrium constant of the reaction.

Temperature Dependence of the Rate of Reaction

From the Arrhenius equation, the following equation can be considered in order to calculate the activation energy of a system:

$$k = Ae\frac{-Ea}{RT}$$

Wherein: k refers to the rate constant, T refers to the absolute temperature, R is the universal gas constant, and A refers to the frequency of correctly oriented collisions. E_a is the activation energy as discussed earlier.

Negative Activation Energy

A negative value of activation energy is found in the Arrhenius equations of reactions wherein the rate of reaction slows down due to an increase in temperature.

In general, reactions which show a negative value for the activation energy are barrier less reactions which involve the capture of molecules in a given potential well.

An increase in the temperature for these reactions results in a reduction in the probability of the colliding molecules to capture each other.

Activation Energies of Endothermic and Exothermic Reactions

Reactions can be broadly classified as endothermic or exothermic reactions based on the role of utilization of thermal energy in these reactions.

Endothermic Reactions

The reactions that require absorption of thermal energy by the reactants to form the necessary products are termed endothermic reactions.

These types of reactions typically have higher activation energies and also require a constant input of energy even after the E_a is provided.

An example of an endothermic reaction is the melting of ice to form water.

Exothermic Reactions

The reactions that release thermal energy during the formation of products are referred to as exothermic reactions.

These types of reactions generally have low activation energies and they do not require a constant input of energy such as in the case of endothermic reactions.

An example of this type of reaction is the combustion of fuel in a bunsen burner. The sparks which need to be provided to start the flame provide the required activation energy to the reacting fuel and the reaction sustains itself after this initial spark is provided.

Zero Order Reactions

A reaction is said to be of zero order if its rate is independent of the concentration of the reactants, i.e., the rate is proportional to the zeroth power of the concentration of the reactants.

For the reaction,

A $\rightarrow$ products, to be of zero order,

$$Rate = -\frac{d[A]}{dt} = k[A]^0$$

d[A] = -kdt

Integrating both sides

[A] = – k t + C

where, C is the constant of integration

At t = o, the concentration of the reactant A = $[A]^o$, where $[A]^o$ is initial concentration of the reactant. Substituting in equation $Rate = -\frac{d[A]}{dt} = k[A]^0$

$$[R]_o = -K \times O + C$$

$$[R]_o = C$$

Substituting the value of C in the equation ([A] = – k t + C)

$$[A] = -kt + [A]_o$$

Comparing equation ($[A] = -kt + [A]_o$) with equation of a straight line.

y = mx + c

If we plot [A] against t, we get a straight line with slope = –k and intercept equal to $[A]_o$.

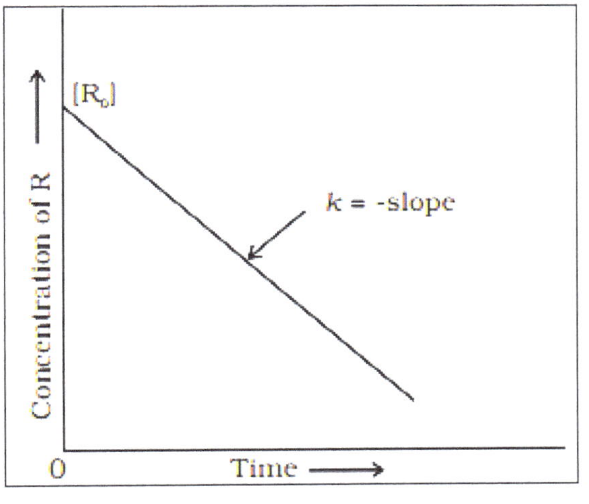

Some photochemical reaction and a few heterogeneous reactions are zero-order reactants. Such reactions are not common.

Examples of Zero Order Reactions

Photochemical Reaction between Hydrogen and Chlorine

$$H_2(g)Cl_2(g)\xrightarrow{hv}2HCl(g)$$

This photochemical reaction is zero-order reaction. The reaction is studied by placing H_2 and Cl_2 gases over water. The rate of reaction is studied by nothing the rate at which water rises in the vessel due to dissociation of HCl formed. the rate of rise of water is the same as the rate of disappearance of H_2 and Cl_2, i.e., the concentration of the gases phase will not change with time, although the quantities will change.

Decomposition of N_2O on hot platinum surface

$$N_2O \rightarrow N_2 + 1/2\ O_2$$

$$\text{Rate } [N_2O]^0 = k[N_2O]^0 = k$$

$$d[N_2O]/dt = k$$

Decomposition of NH_3 in presence of molybdenum or tungsten is a zero-order reaction

$$2NH_3 \xrightarrow{Mo} N_2 + 3H_2$$

The surface of the catalyst is almost completely covered by NH_3 molecules. The adsorption of gas on the surface cannot change by increasing the pressure or concentration of NH_3. Thus, the concentration of gas phase remains constant although the product is formed. Therefore, this reaction zero order kinetics.

Other examples of zero order are,

Iodations of acetone in presence of H+ ions

$$CH_3COCH + I_2 \xrightarrow{H^+} ICH_2COCH_3 + HI$$

The rate equation of this reaction does not include $[I_2]$ factor, i.e.,

$$-dx/dt = k[CH_3COCH_3][H^+]$$

Characteristics of Zero Order Reaction

- The concentration of reactant decreases linearly with time.$[A]_t = [A]_o - kt$

- The time required for the reaction to be complete, i.e., time at which [A] is zero. $t_{completion} = [A]_0/k$ = (Initial concentration)/(Rate constant)
- The units of k are mol L^{-1} $time^{-1}$.

Half-Time or Half-Life Period of a Zero Order Reaction

The half-time of a reaction is defined as the time required to reduce the concentration of the reactant to half of its initial value. It is denoted by the symbol $t_{1/2}$. Thus,

For zero order reaction, we know that

$$[A] = -kt + [A]_0$$

$$k = \frac{[A]_o - [A]}{t}$$

at half life of reaction,

$$t = t_{½} \ \& \ [A] = [A]_o / 2$$

So

$$k = \frac{[A]_o - [A]_o / 2}{t} = \frac{[A]_o}{2t_{1/2}}$$

$$\Rightarrow t_{1/2} = \frac{[A]_o}{2k}$$

First Order Reaction

A reaction is said to be first order if its rate is determined by the change of one concentration term only.

One can say that a first order reaction is one whose rate varies as 1st power of the concentration of the reactant i.e. the rate increases as number of times as the number of times the concentration of reactant is increased.

Consider the reaction

$$A \rightarrow \text{products}$$

Let

$[A]_0$ = Initaial Concentration of A

$[A]_t$ = The concentration of A after time t

For the reaction to be of first order,

$$Rate = -\frac{d[A]]}{dt} = k[A]$$

$$\Rightarrow \frac{d[A]}{[A]} = -kdt$$

Intigrating this equation,

$$\int_{[A]_o}^{[A]} \frac{d[A]}{[A]} = -k\int_0^t dt$$

We get

$$\ln [A] - \ln[A]_o = -K(t-o)$$

$$ln\frac{[A]}{[A]_o} = -kt$$

$$\Rightarrow ln\frac{[A]_o}{[A]} = kt$$

$$\Rightarrow log\frac{[A]_o}{[A]} = \frac{kt}{2.303}$$

$$\Rightarrow k = \frac{1}{t}ln\frac{[A]_o}{[A]}$$

On comparing the equation number ($ln\frac{[A]_o}{[A]} = kt$) with equation of straight line i.e. y=mx+c we come to know that , If we plot a graph between log $[A]_o/[A]$ vs t, the slope of line= k/2.303

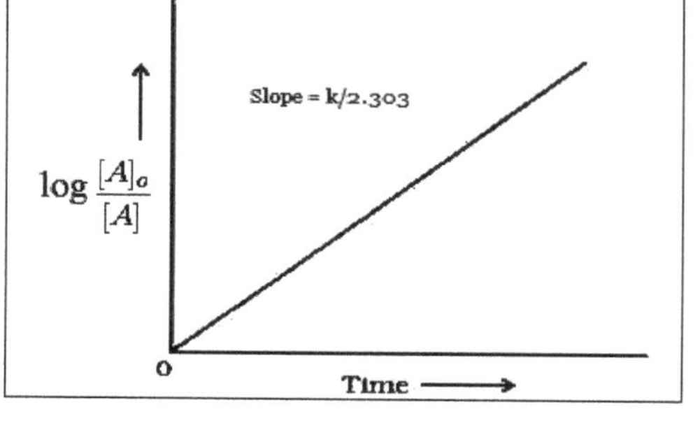

Unit of Rate Constant for First Order Reaction (k) = min−1 or s-1

- The rate constant of a first order reaction has only time unit. It has no concentration unit.
- This means the numerical value of k for a first order reaction is independent of the unit in which concentration is expressed.
- If concentration unit is changed the numerical value of k for a first order reaction will not change.
- However, it would change with change in time unit. Say, k is 6.0×10^{-3} min then it may also be written as $1 \times 10^{-4}\ s^{-1}$ i.e. numerical value of k will decrease 60 times if time unit is changed from however to minute or from minute to second.

Half-Time or Half-life Period of a First Order Reaction

The half-time of a reaction is defined as the time required to reduce the concentration of the reactant to half of its initial value. It is denoted by the symbol $t_{1/2}$. Thus,

For first order reaction, we know that

$$k = \frac{1}{t} ln \frac{[A]_o}{[A]}$$

at half life of reaction,

$$t = t_{1/2}\ \&\ [A] = [A]_0/2$$

So,

$$k = \frac{1}{t_{1/2}} ln \frac{[A]_o}{\frac{[A]_o}{2}}$$

$$\Rightarrow t_{1/2} = \frac{0.693}{k}$$

- Since k is a constant for a given reaction at a given temperature and the expression lacks any concentration term so half-time of a 1st order reaction is a constant independent of initial concentration of reactant.
- This means if we start with 4 mole L^{-1} of a reactant reacting by first-order kinetics and after 20 minute it is reduced to 2 mole L^{-1} will also be 20 minute.
- That is, after 20 minutes from the start of reaction the concentration of the reactant will be 2 mole L^{-1}, after 40 minutes from the start of reaction of concentration is 1 mole L^{-1}. After 60 minutes from the start of reaction the

concentration of the reactant will be reduced to 0.5 mole L^{-1}. In other words, if during 20 minute 50% of the reaction completes, then in 40 minute 75%, in 60 minute 85.5% of the reaction and on will complete as shown with following plot.

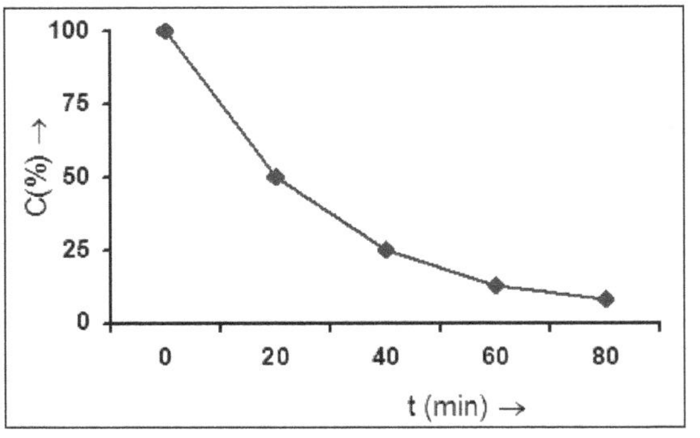

Examples of First Order Reactions

- Decomposition of H_2O_2 in aqueous solution

$$H_2O_2 \rightarrow H_2O + 1/2\ O_2$$

- Hydrolysis of methyl acetate in presence of mineral acids.

$$CH_3COOCH_3 + H_2O \rightarrow CH_3COOH + CH_3OH$$

- Inversion of cane sugar in presence of mineral acids.

$$C_{12}H_{22}O_{11} + H_2O \xrightarrow{H^+} C_6H_{12}O_6 + C_6H_{12}O_6$$

- Decomposition of ammonium nitrite in aqueous solution.

$$NH_4NO_2 \rightarrow N_2 + 2H_2O$$

- Hydrolysis of diazo derivatives.

$$C_5H_5N = NCl + H_2O \rightarrow C_6H_5OH + N_2 + HCl$$

In case of gases, pressure can be used in place of concentration.

Some Simple First Order Reactions

As we have to generally deal with only first order reactions, we examine some of these reactions from the point of calculating the rate constant based on different experimental data.

Now we present several problems in which we shall learn how to calculate the rate constant of reactions based on the variety of data given.

- We are given a first order reaction $A \rightarrow B+C$ where we assume that A, B and C are gases. The data given to us is

  $A \rightarrow B + C$

Time	0	t
Partial pressure of A	P_1	P_2

And we have to find the rate constant of the reaction.

Since A is a gas and assuming it to be ideal, we can state that $P_A = [A]RT$

[From PV = nRT]. \ at t = 0, $P_1 = [A]_0$ RT and at t = t, $P_2 = [A]_t RT$. Thus the ratio of the concentration of A at two different time intervals is equal to the ratio of its partial pressure at those same time intervals.

$$\therefore \frac{[A]_0}{[A]_t} = \frac{P_1}{P_2}$$

$$\therefore k = \frac{1}{t} \text{In} \frac{P_1}{P_2}$$

- Now we consider the same reaction with different set of data $A \rightarrow B + C$

Time	0	t
Total pressure of A +B+C	P_1	P_2

Find k

In this we are given total pressure of the system at these time intervals. The total pressure obviously includes the pressure of A, B and C. At t = 0, the system would only have A. Therefore the total pressure at t = 0 would be the initial pressure of A \ P1 is the initial pressure of A. At time t let us assume that some moles of A decomposed to give B and C because of which its pressure is reduced by an amount x while that of B and C is increased by x each. That is,

	$A \rightarrow$	$B +$	C
Intial	P_1	0	0
At time t	$P_1 - x$	x	x

$\therefore$ total pressure at time t = $P_1 - x = P_2$

$$\Rightarrow x = P_2 - P_1$$

Now the pressure of A at time t would be $P_1 - x = P_1 - (P_2 - P_1) = 2P_1 - P_2$

$$\therefore \mathrm{k} = \mathrm{In}\frac{[A]_0}{[A]_t} = \mathrm{In}\frac{\mathrm{P_1}}{(2\mathrm{P_1} - \mathrm{P_2})}$$

- Now let us move on to the next case. In this case we have A → B + C

Time	t	∞
Total pressure of A+B+C	P_2	P_3

Find k

Here ∞ means that the reaction is complete. Now we have

$$\begin{array}{llll} & A & \rightarrow B & + \quad C \\ At\ t=0 & P_1 & 0 & 0 \\ At\ t=t & (P_1 - x) & x & x \\ At\ t=\infty & 0 & P_1 & P_1 \end{array}$$

$$\therefore 2\mathrm{P_1} = \mathrm{P_3}$$

$$\Rightarrow \mathrm{P_1} = \frac{\mathrm{P_3}}{2}$$

At time t,

$$P_1 + x = P_2$$

$$\Rightarrow \frac{P_3}{2} + x = P_2 \Rightarrow x = P_2 - \frac{P_3}{2}$$

$$\Rightarrow P_1 - x = \frac{P_3}{2} - \left(P_2 - \frac{P_3}{2}\right) = P_3 - P_2$$

$$\therefore k = \frac{1}{t}\mathrm{In}\frac{[A]_0}{[A]_t} = \frac{1}{t}\mathrm{In}\frac{p_3/2}{(P_3/P_2)} = \frac{1}{t}\mathrm{In}\frac{P_3}{(P_3 - P_2)}$$

- Now let us assume that A, B and C are substances present in a solution.

  From a solution of A, a certain amount of the solution (small amount) is taken and titrated with a suitable reagent that reacts with A. The volume of the reagent used is V_1 at t = o and V_2 at t = t

Time	o	T
Volume of reagent	V_1	V_2

The reagent reacts only with A. Find k.

It can be understood that the volume of the reagent consumed is directly proportional to the concentration of A. Therefore the ratio of volume of the reagent consumed against A at t=0 and t = t is equal to $[A]_o / [A]$

$$\therefore k = \frac{1}{t}\text{In}\frac{[A]_0}{[A]_t} = \frac{1}{t}\text{In}\frac{V_1}{V_2}$$

- Now consider this situation $A \rightarrow B + C$

Time	T	∞
Volume of reagent	V_2	V_3

Reagent reacts with all A, B and C and have 'n' factors in the ratio of 1:2:3 with the reagent.

Find k

$$\begin{array}{llll} & A & \rightarrow B & + \quad C \\ At\ t=0 & V_1 & 0 & 0 \\ At\ t=t & V_1 - x & 2x & 3x \\ At\ t=\infty & 0 & P_1 & 3V_1 \end{array}$$

$$\therefore 5V_1 = V_3$$

$$V_1 = V_3/5$$

$$V_1 + 4\ x = V_2$$

$$x = \frac{V_2 - V_1}{4}$$

$$\therefore V_1 - x = \frac{V_3}{5} - \frac{V_2}{4} + \frac{V_3}{20} = \frac{5V_3 - 5V_2}{20}$$

$$\therefore k = \frac{1}{t}\text{In}\frac{[A]_0}{[A]_t} = \frac{1}{t}\text{In}\frac{V_3/5}{5V_3 - 5V_2/20}$$

$$k = \frac{1}{t}\text{In}\frac{4V_3}{5(V_3 - V_2)}$$

Now, we will consider a reaction $A \xrightarrow{D} B + C$ which is catalysed by D.

We will assume in this problem that the concentration of the catalyst remains constant throughout.

The data given to us is

Time	0	t	∞
Volume of reagent	V_1	V_2	V_3

The reagent reacts with all (A, B, C and D). Assume that 'n' factor of A,B and C are in the ratio of 1:1:1 and that of D is not known. Find k.

Let V_A be the volume of the reagent required by A initially and V_D be the volume required by D.

$$\backslash V_1 = V_A + V_D$$

$V_2 = (V_A - x) + x + x + V_D$ (x is the volume of the reagent required for those moles of A that have reacted to give B and C).

$$V_3 = 2V_A + V_D$$

We can see that, $V_3 - V_1 = V_A$

$$\text{And } V_3 - V_2 = V_A - x$$

$$\therefore k = \frac{1}{t} \ln \frac{[A]_0}{[A]_t} = \frac{1}{t} \ln \frac{(V_3 - V_1)}{(V_3 - V_2)}$$

- Now we shall see how to find the rate constant of a reaction using a very different set of data.

  There are some organic compounds which have a property of rotating a plane polarized light in a particular direction by a particular value. The compounds are called optically active compounds. One reaction in which an optically active substance converts to some other optically active substance is,

  $$\text{Sucrose} \xrightarrow{H^+} \text{Glucose} + \text{Fructose}$$

  Sucrose, Glucose and Fructose are all optically active and while the first two compounds are dextro rotatory (rotating the plane polarised light in the right hand direction and the last is laevo rotatory (rotating the plane polarized light in the left hand direction). All the three compounds rotate the plane polarised light by different angles and their rotation is directly proportional to concentration.

Now the problem is $S \rightarrow G + F$ and the data is

Time	o	t
Rotation of sucrose	r_o	r_t

Find k.

Let the rotation of Sucrose be r_1^o per mole and the initial moles of Sucrose be a.

$$r_o = ar_1^o$$

Let the moles of Sucrose that is converted to Glucose and Fructose be x.

$$\therefore r_t = (a - x) r_1^o$$

$$\therefore \frac{a}{a - x} = \frac{r_0}{r_t}$$

$$\therefore k = \frac{1}{t} \ln \frac{a}{a - x} = \frac{1}{t} \ln \frac{r_0}{r_t}$$

Pseudo First Order Reaction

We know that order of reaction depends on the dependency of the rate of reaction on the concentration of reactants. That is if the rate is independent of the concentrations of reactants, the order of reaction is zero. Similarly, if the rate of reaction is proportional to the first power of the concentration of the reactant, then the order of reaction is one.

But sometimes the order of a reaction can be made to alter by changing the concentration of the reactants, that is, increasing or decreasing the concentration of one or the other reactant.

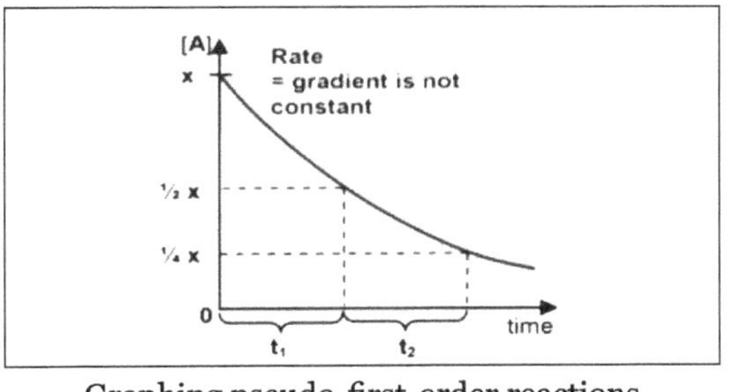

Graphing pseudo-first-order reactions

A reaction which is not first-order reaction naturally but made first order by increasing or decreasing the concentration of one or the other reactant is known as Pseudo first order reaction. Pseudo means 'fake'.

Therefore, it is clear from the name itself that it is not first-order reaction by nature. The order of reaction is made one by altering certain conditions. Let's understand more about pseudo-first order reactions with some examples.

Examples of Pseudo First Order Reaction

- Consider a reaction in which one reactant is in excess, say hydrolysis of ethyl acetate. During the hydrolysis of 0.01 mol of ethyl acetate with 10 mol of water, the reaction goes like this,

$$CH_3COOC_2H_5 + H_2O \rightarrow CH_3COOH + C_2H_5OH$$

Time	$CH_3COOC_2H_5$	H_2O	CH_3COOH	C_2H_5OH
t = 0	0.01 mol	10 mol	0 mol	0 mol
t	0 mol	9.9 mol	0.01 mol	0.01 mol

As the concentration of water does not get altered much during the process (because it is in excess), we can take it as constant.

So in the rate equation,

Rate = k′ $[CH_3COOC_2H_5]$ $[H_2O]$ the term $[H_2O]$ can be taken constant.

Thus, the equation becomes,

$$\text{Rate} = k\,[CH_3COOC_2H_5] \text{ where } k = k'[H_2O]$$

Therefore, the order of reaction now becomes one, that is the reaction is now first order reaction. Such reactions are called pseudo-first order reactions.

- Another example of pseudo first-order reaction is the inversion of cane sugar.

  $$C_{12}H_{22}O_{11} + H_2O \rightarrow C_6\,H_{12}O_6 + C_6\,H_{12}O_6$$

  Cane sugar Glucose Fructose

  $$Rate = k\left[C_{12}H_{22}O_{11}\right]$$

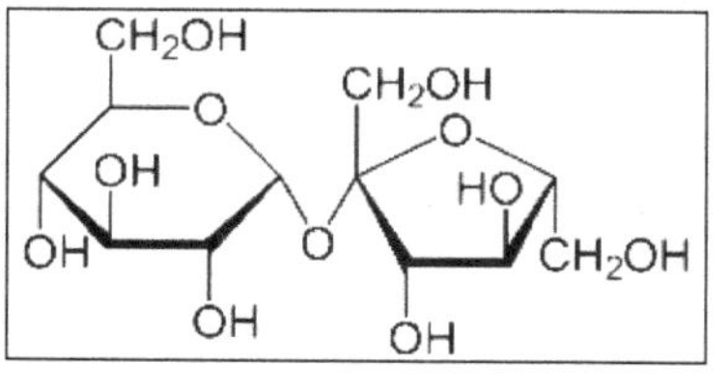

In pseudo-first order reactions, we are basically isolating a reactant by increasing the concentration of the other reactants. When the other reactants are in excess, change in their concentrations does not affect the reaction much, Therefore, now the reaction only depends on the concentration of the isolated reactant. The concentrations of all the other reactants are taken as constant in the rate law. Thus, the order of reaction becomes one.

Difference between Pseudo First Order Reaction and First Order Reaction

The main difference is obviously that, in a first order reaction, the order of reaction is one by nature. A pseudo first-order reaction is second order reaction by nature but has been altered to make it a first order reaction.

The second difference is that in a first order reaction, the rate of reaction depends on all the reactants whereas, in a pseudo-first-order reaction, the rate of reaction depends only on the isolated reactant since a difference in concentration of the reactant in excess will not affect the reaction.

Second Order Reaction

Second order reactions can be defined as chemical reactions wherein the sum of the exponents in the corresponding rate law of the chemical reaction is equal to two. The rate of such a reaction can be written either as $r = k[A]^2$, or as $r = k[A][B]$.

From the rate law equations given above, it can be understood that second order reactions are chemical reactions which depend on either the concentrations of two first-order reactants or the concentration of one-second order reactant.

Since second order reactions can be of the two types described above, the rate of these reactions can be generalized as follows:

$$r = k[A]^x [B]^y$$

Where the sum of x and y (which corresponds to the order of the chemical reaction in question) equals two.

Examples of Second Order Reactions

A few examples of second order reactions are given below:

$$H^+ + OH^- \rightarrow H_2OC + O_2 \rightarrow CO + O$$

The two examples given above are the second order reactions depending on the concentration of two separate first order reactants.

$$2NO_2 \rightarrow 2NO + O_2\ 2HI \rightarrow I_2 + H_2$$

These reactions involve one second order reactant yielding the product.

Differential and Integrated Rate Equation for Second Order Reactions

Considering the scenario where one second order reactant forms a given product in a chemical reaction, the differential rate law equation can be written as follows:

$$-\frac{d[R]}{dt} = k[R]^2$$

In order to obtain the integrated rate equation, this differential form must be rearranged as follows:

$$\frac{d[R]}{[R]^2} = -kdt$$

Now, integrating on both sides in consideration of the change in the concentration of reactant between time o and time t, the following equation is attained.

$$\int \frac{[R]t}{[R]0} \frac{d[R]}{[R]^2} = -k \int_0^t dt$$

From the power rule of integration, we have:

$$\int \frac{dx}{x^2} = -\frac{1}{x} + C$$

Where C is the constant of Integration. Now, using this power rule in the previous equation, the following equation can be attained.

$$\frac{1}{[R]_t} - \frac{1}{[R]_0} = kt$$

Which is the required integrated rate expression of second order reactions.

Graph of a Second Order Reaction

Generalizing $[R]_t$ as [R] and rearranging the integrated rate law equation of reactions of the second order, the following reaction is obtained.

$$\frac{1}{[R]} = kt + \frac{1}{[R]_0}$$

Plotting a straight line (y=mx + c) corresponding to this equation (y = 1/[R] , x = t , m = k , c = $1/[R]_0$)

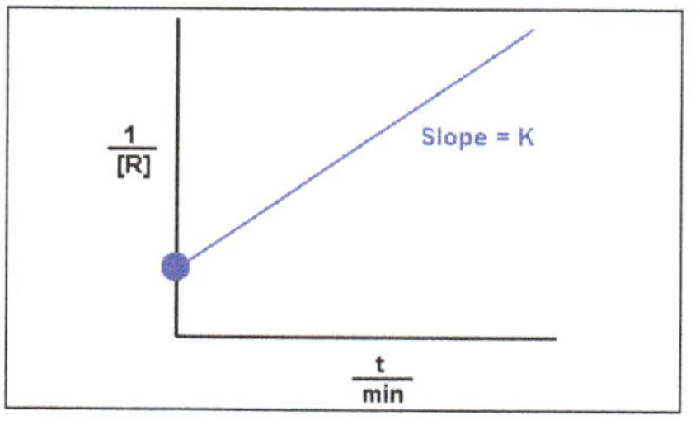

It can be observed that the slope of the straight line is equal to the value of the rate constant, k.

Half-Life of Second-Order Reactions

The half-life of a chemical reaction is the time taken for half of the initial amount of reactant to undergo the reaction.

Therefore, while attempting to calculate the half life of a reaction, the following substitutions must be made:

$$[R] = \frac{[R]_0}{2}$$

$$\text{And, } t = t_{1/2}$$

Now, substituting these values in the integral form of the rate equation of second order reactions, we get:

$$\frac{1}{\frac{[R]_0}{2}} - \frac{1}{[R]_0} = kt_{1/2},$$

Therefore, the required equation for the half life of second order reactions can be written as follows.

$$t_{1/2} = \frac{1}{k[R]_0}$$

This equation for the half life implies that the half life is inversely proportional to the concentration of the reactants.

Carbon 14 Dating

Carbon-14 dating is also called radiocarbon dating, method of age determination that depends upon the decay to nitrogen of radiocarbon (carbon-14). Carbon-14 is continually formed in nature by the interaction of neutrons with nitrogen-14 in the Earth's atmosphere; the neutrons required for this reaction are produced by cosmic rays interacting with the atmosphere.

Radiocarbon present in molecules of atmospheric carbon dioxide enters the biological carbon cycle: it is absorbed from the air by green plants and then passed on to animals through the food chain. Radiocarbon decays slowly in a living organism, and the amount lost is continually replenished as long as the organism takes in air or food. Once the organism dies, however, it ceases to absorb carbon-14, so that the amount of the radiocarbon in its tissues steadily decreases. Carbon-14 has a half-life of 5,730 ± 40 years—i.e., half the amount of the radioisotope present at any given time will undergo spontaneous disintegration during the succeeding 5,730 years. Because carbon-14 decays at this constant rate, an estimate of the date at which an organism died can be made by measuring the amount of its residual radiocarbon.

The carbon-14 method was developed by the American physicist Willard F. Libby about 1946. It has proved to be a versatile technique of dating fossils and archaeological specimens from 500 to 50,000 years old. The method is widely used by Pleistocene geologists, anthropologists, archaeologists, and investigators in related fields.

Radiocarbon dating is a method that provides objective age estimates for carbon-based materials that originated from living organisms. 1 An age could be estimated by measuring the amount of carbon-14 present in the sample and comparing this against an internationally used reference standard.

The impact of the radiocarbon dating technique on modern man has made it one of the most significant discoveries of the 20th century. No other scientific method has managed to revolutionize man's understanding not only of his present but also of events that already happened thousands of years ago. Archaeology and other human sciences use radiocarbon dating to prove or disprove theories. Over the years, carbon 14 dating has also found applications in geology, hydrology, geophysics, atmospheric science, oceanography, paleoclimatology and even biomedicine.

References

- Chemical-kinetics, science: britannica.com, Retrieved 4 March, 2019
- Definition-of-chemical-kinetics: thoughtco.com, Retrieved 6 May, 2019
- Reaction_Mechanisms, Rate_Laws, Kinetics, Physical_and_Theoretical_Chemistry: chem.libretexts.org, Retrieved 16 February, 2019
- Collision-theory-of-chemical-reactions, chemical-kinetics, chemistry: toppr.com, Retrieved 28 June, 2019
- Activation-energy-definition: thoughtco.com, Retrieved 23 August, 2019
- Activation-energy, chemistry: byjus.com, Retrieved 21 April, 2019
- Zero-order-reaction, chemical-kinetics, physical-chemistry: askiitians.com, Retrieved 10 July, 2019
- First-order-reaction, chemical-kinetics, physical-chemistry: askiitians.com, Retrieved February 19, 2019
- Pseudo-first-order-reaction, chemical-kinetics, chemistry: toppr.com, Retrieved 20 January, 2019
- Second-order-reaction, chemistry: byjus.com, Retrieved 18 March, 2019
- Carbon-14-dating, science: britannica.com, Retrieved 14May, 2019
- Carbon-dating: radiocarbon.com, Retrieved 25 April, 2019

5

Chemical Thermodynamics

The study of the interrelation of heat and work with chemical reactions is known as chemical thermodynamics. In order to study chemical questions and the spontaneity of processes it involves laboratory measurements of thermodynamic properties and diverse mathematical methods. The diverse aspects of chemical thermodynamics in the current scenario have been thoroughly discussed in this chapter.

Thermodynamics is science of the, relationship between heat, work, temperature, and energy. In broad terms, thermodynamics deals with the transfer of energy from one place to another and from one form to another. The key concept is that heat is a form of energy corresponding to a definite amount of mechanical work.

Heat was not formally recognized as a form of energy until about 1798, when Count Rumford (Sir Benjamin Thompson), a British military engineer, noticed that limitless amounts of heat could be generated in the boring of cannon barrels and that the amount of heat generated is proportional to the work done in turning a blunt boring tool. Rumford's observation of the proportionality between heat generated and work done lies at the foundation of thermodynamics.

Laws of Thermodynamics

Zeroth Law of Thermodynamics

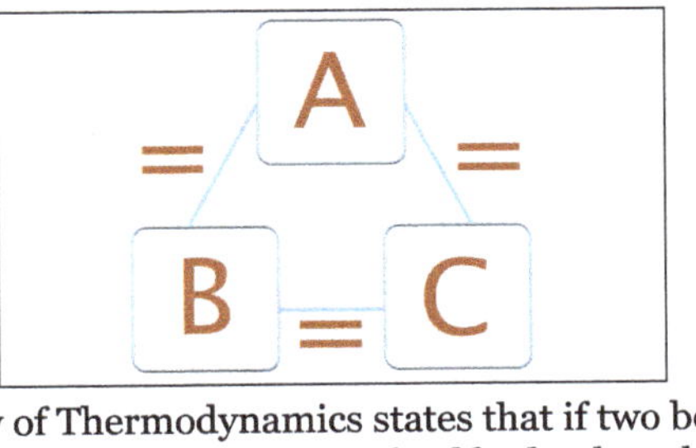

The Zeroth Law of Thermodynamics states that if two bodies are each in thermal equilibrium with some third body, then they are also in equilibrium with each other.

The Zeroth Law of Thermodynamics states that if two bodies are each in thermal equilibrium with some third body, then they are also in equilibrium with each other. Thermal equilibrium means that when two bodies are brought into contact with each other and separated by a barrier that is permeable to heat, there will be no transfer of heat from one to the other.

This says in essence that the three bodies are all the same temperature. James Clerk Maxwell put this perhaps more simply when he said, "All heat is of the same kind." What is most important is that the Zeroth Law establishes that temperature is a fundamental and measurable property of matter.

First Law of Thermodynamics

The First Law of Thermodynamics states that heat is a form of energy, and thermodynamic processes are therefore subject to the principle of conservation of energy. This means that heat energy cannot be created or destroyed. It can, however, be transferred from one location to another and converted to and from other forms of energy.

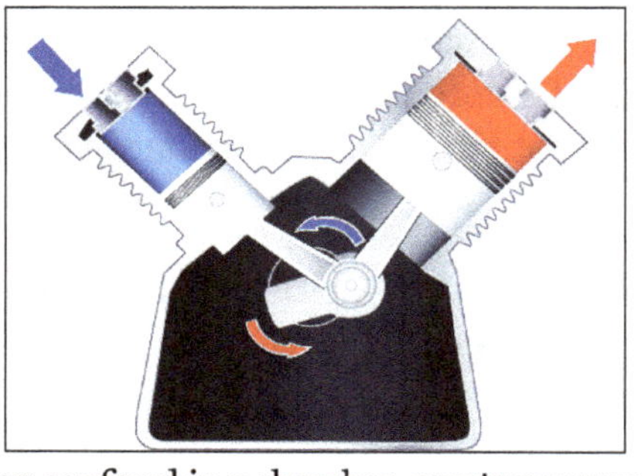

A hot gas, when confined in a chamber, exerts pressure on a piston, causing it to move downward. The movement can be harnessed to do work equal to the total force applied to the top of the piston times the distance that the piston moves.

Heat Engines

The most common practical application of the First Law is the heat engine. Heat engines convert thermal energy into mechanical energy and vice versa. Most heat engines fall into the category of open systems. The basic principle of a heat engine exploits the relationships among heat, volume and pressure of a working fluid. This fluid is typically a gas, but in some cases it may undergo phase changes from gas to liquid and back to a gas during a cycle.

When gas is heated, it expands; however, when that gas is confined, it increases in pressure. If the bottom wall of the confinement chamber is the top of a movable piston, this pressure exerts a force on the surface of the piston causing it to move downward. This movement can then be harnessed to do work equal to the total force applied to the top of the piston times the distance that the piston moves.

There are numerous variations on the basic heat engine. For instance, steam engines rely on external combustion to heat a boiler tank containing the working fluid, typically water. The water is converted to steam, and the pressure is then used to drive a piston that converts heat energy to mechanical energy. Automobile engines, however, use internal combustion, where liquid fuel is vaporized, mixed with air and ignited inside a cylinder above a movable piston driving it downward.

Refrigerators, Air Conditioners and Heat Pumps

Refrigerators and heat pumps are heat engines that convert mechanical energy to heat. Most of these fall into the category of closed systems. When a gas is compressed, its temperature increases. This hot gas can then transfer heat to its surrounding environment. Then, when the compressed gas is allowed to expand, its temperature becomes colder than it was before it was compressed because some of its heat energy was removed during the hot cycle. This cold gas can then absorb heat energy from its environment. This is the working principal behind an air conditioner. Air conditioners don't actually produce cold; they remove heat. The working fluid is transferred outdoors by a mechanical pump where it is heated by compression. Next, it transfers that heat to the outdoor environment, usually through an air-cooled heat exchanger. Then, it is brought back indoors, where it is allowed to expand and cool so it can absorb heat from the indoor air through another heat exchanger.

A heat pump is simply an air conditioner run in reverse. The heat from the compressed working fluid is used to warm the building. It is then transferred outside where it expands and becomes cold, thereby allowing it to absorb heat from the outside air, which even in winter is usually warmer than the cold working fluid.

Second Law of Thermodynamics

The Second Law of Thermodynamics states that the state of entropy of the entire universe, as an isolated system, will always increase over time. The second law also states that the changes in the entropy in the universe can never be negative.

Derivation and Explanation

To understand why entropy increases and decreases, it is important to recognize that two changes in entropy have to considered at all times. The entropy change of the surroundings and the entropy change of the system itself. Given the entropy change of the universe is equivalent to the sums of the changes in entropy of the system and surroundings:

$$\Delta S_{univ} = \Delta S_{sys} + \Delta S_{surr} = \frac{q_{sys}}{T} + \frac{q_{surr}}{T}$$

In an isothermal reversible expansion, the heat q absorbed by the system from the surroundings is,

$$q_{rev} = nRT \ln \frac{V_2}{V_1}$$

Since the heat absorbed by the system is the amount lost by the surroundings, $q_{sys} = -q_{surr}$. Therefore, for a truly reversible process, the entropy change is

$$\Delta S_{univ} = \frac{nRT\ln\frac{V_2}{V_1}}{T} + \frac{-nRT\ln\frac{V_2}{V_1}}{T} = 0$$

If the process is irreversible however, the entropy change is

$$\Delta S_{univ} = \frac{nRT\ln\frac{V_2}{V_1}}{T} > 0$$

If we put the two equations for $\ddot{A}S_{univ}$ together for both types of processes, we are left with the second law of thermodynamics,

$$\Delta S_{univ} = \Delta S_{sys} + \Delta S_{surr} \geq 0$$

Where $\ddot{A}S_{univ}$ equals zero for a truly reversible process and is greater than zero for an irreversible process. In reality, however, truly reversible processes never happen (or will take an infinitely long time to happen), so it is safe to say all thermodynamic processes we encounter everyday are irreversible in the direction they occur.

The second law of thermodynamics can also be stated that "all spontaneous processes produce an increase in the entropy of the universe"

Third Law of Thermodynamics

The third law of thermodynamics states that the entropy of a perfect crystal at a temperature of zero Kelvin (absolute zero) is equal to zero.

Entropy, denoted by 'S', is a measure of the disorder/randomness in a closed system. It is directly related to the number of microstates (a fixed microscopic state that can be occupied by a system) accessible by the system, i.e. the greater the number of microstates the closed system can occupy, the greater its entropy. The microstate in which the energy of the system is at its minimum is called the ground state of the system.

At a temperature of zero Kelvin, the following phenomena can be observed in a closed system:

- The system does not contain any heat.
- All the atoms and molecules in the system are at their lowest energy points.

Therefore, a system at absolute zero has only one accessible microstate – it's ground state. As per the third law of thermodynamics, the entropy of such a system is exactly zero.

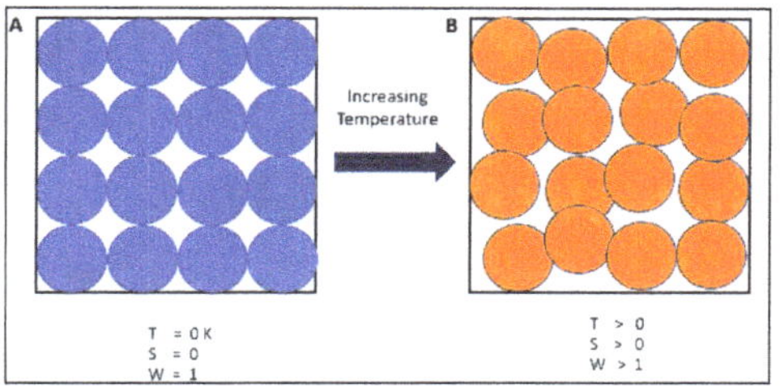

This law was developed by the German chemist Walther Nernst between the years 1906 and 1912.

The Nernst statement of the third law of thermodynamics implies that it is not possible for a process to bring the entropy of a given system to zero in a finite number of operations.

The American physical chemists Merle Randall and Gilbert Lewis stated this law in a different manner: when the entropy of each and every element (in their perfectly crystalline states) is taken as 0 at absolute zero temperature, the entropy of every substance must have a positive, finite value. However, the entropy at absolute zero can be equal to zero, as is the case when a perfect crystal is considered.

The Nernst-Simon statement of the 3rd law of thermodynamics can be written as: for a condensed system undergoing an isothermal process that is reversible in nature, the associated entropy change approaches zero as the associated temperature approaches zero.

Another implication of the third law of thermodynamics is: the exchange of energy between two thermodynamic systems (whose composite constitutes an isolated system) is bounded.

Why is it Impossible to Achieve a Temperature of Zero Kelvin?

For an isentropic process that reduces the temperature of some substance by modifying some parameter X to bring about a change from 'X_2' to 'X_1', an infinite number of steps must be performed in order to cool the substance to zero Kelvin.

This is because the third law of thermodynamics states that the entropy change at absolute zero temperatures is zero. The entropy v/s temperature graph for any isentropic process attempting to cool a substance to absolute zero is illustrated below.

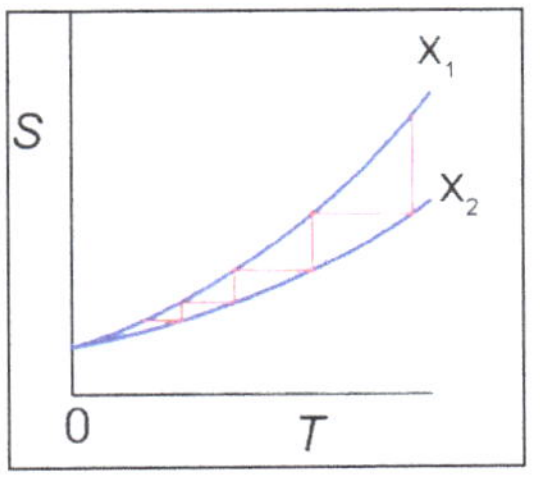

From the graph, it can be observed that – the lower the temperature associated with the substance, the greater the number of steps required to cool the substance further.

As the temperature approaches zero kelvin, the number of steps required to cool the substance further approaches infinity.

Mathematical Explanation of the Third Law

As per statistical mechanics, the entropy of a system can be expressed via the following equation:

$$S - S_o = k_B \ln\Omega$$

Where,

- S is the entropy of the system.
- S_o is the initial entropy.
- k_B denotes the Boltzmann constant.
- **Ω** refers to the total number of microstates that are consistent with the system's macroscopic configuration.

Now, for a perfect crystal that has exactly one unique ground state, **Ω** = 1. Therefore, the equation can be rewritten as follows:

$$S - S_o = k_B \ln(1) = 0 \text{ [because } \ln(1) = 0]$$

When the initial entropy of the system is selected as zero, the following value of 'S' can be obtained:

$$S - 0 = 0 \Rightarrow S = 0$$

Thus, the entropy of a perfect crystal at absolute zero is zero.

Applications of the Third Law of Thermodynamics

An important application of the third law of thermodynamics is that it helps in the calculation of the absolute entropy of a substance at any temperature 'T'. These determinations are based on the heat capacity measurements of the substance. For any solid, let So be the entropy at 0 K and S be the entropy at T K, then

$$\Delta S = S - S_0 = \int_0^T \frac{C_p d}{T}$$

According to the third law of thermodynamics, S_o= 0 at 0 K,

$$S = \int_0^T \frac{C_p}{T} dT$$

The value of this integral can be obtained by plotting the graph of C_p/T versus T and then finding the area of this curve from 0 to T. The simplified expression for the absolute entropy of a solid at temperature T is as follows:

$$S = \int_0^T \frac{C_p}{T} dT = \int_0^T C_p \, d \ln T$$
$$= C_p \ln T = 2.303\ C_p \log T$$

Here C_p is the heat capacity of the substance at constant pressure and this value is assumed to be constant in the range of 0 to T K.

To learn more about the third law of thermodynamics and other laws of thermodynamics, register with BYJU'S and download the mobile application on your smartphone.

Heat, Energy and Work

Heat and work are both forms of energy. They are also related forms, in that one can be transformed into the other. Heat energy (such as steam engines) can be used to do work (such as pushing a train down the track). Work can be transformed into heat, such as might be experienced by rubbing your hands together to warm them up.

Work and heat can both be described using the same unit of measure. Sometimes the calorie is the unit of measure, and refers to the amount of heat required to raise one (1) gram of water one (1) degree Celsius. Heat energy is measured in kilocalories, or 1000 calories. Typically, we use the SI units of Joules (J) and kilojoules (kJ). One calorie of heat is equivalent to 4.187 J. You will also encounter the term specific heat, the heat required to raise one (1) gram of a material one (1) degree Celsius. Specific heat, given by the symbol "C", is generally defined as:

$$C = \frac{q}{M \Delta T}$$

Where,

C = specific heat in cal/g-°C

q = heat added in calories,

m = mass in grams

ΔT = rise in temperature of the material in °C.

The value of C for water is 1.00 cal/g-°C.

The values for specific heat that are reported in the literature are usually listed at a specific pressure and volume, and you need to pay attention to these settings when using values from textbooks in problems or computer models.

Example Problem: If a 2.34 g substance at 22 °C with a specific heat of 3.88 cal/g- °C is heated with 124 cal of energy, what is the new temperature of the substance?

Solution,

$$\Delta T = \frac{q}{MC}$$

$$\Delta T = \frac{(124)}{(2.34)(3.88)} = 13.7\ °C$$

$$\text{new T} = 22 + 13.7 = 35.7\ °C$$

Two other common heat variables are the heat of fusion and the heat of vaporization. Heat of fusion is the heat required to melt a substance at its melting temperature, while the heat of vaporization is the heat required to evaporate the substance at its boiling point.

Chemical work is primarily related to that of expansion. In physics, work is defined as:

$$w = d \times f$$

Where:

w = work, in joules (N×m) (or calories, but we are using primarily SI units)

d = distance in meters

f = opposing force in Newtons (kg×m/s2)

In chemical reactions, work is generally defined as,

w = distance × (area × pressure)

The value of distance times area is actually the volume. If we imagine a reaction taking place in a container of some volume, we measure work by pressure times the change in volume.

$$w = \Delta V \times P$$

Where,

ΔV is the change in volume, in liters

If $\Delta V = 0$, then no work is done.

Example Problem: Calculate the work that must be done at standard temperature and pressure (STP is 0 °C and 1 atm) to make room for the products of the octane combustion:

$$2\ C_8H_{18} + 25\ O_2 \rightarrow 16\ CO_2 + 18\ H_2O$$

Solution

Knowing that 25 moles of gas are replaced by 34 moles of gas in this reaction, we can calculate a net increase of 9 moles of gas. Knowing the molar volume of an ideal gas at STP (22.4 L/mol), the change in volume and the work of expansion can be calculated

$$dV = 9 \text{ moles} \times 22.4 \text{ L/mol} = 202 \text{ L}$$

The external pressure is 1.0 atm (standard pressure), so the work required is:

$$w = dV \times P = 202 \text{ L} \times 1.00 \text{ atm} = 202 \text{ l-atm}$$

Using the conversion factor of 1 L-atm = 101 J, the amount of work in joules is:

$$w = 202 \text{ L-atm} \times 101 \text{ j/L-atm} = 2000 \text{ J, or 2kJ of energy}$$

Energy

You might remember the first law of thermodynamics: energy cannot be created or destroyed. Energy can only change form. Chemically, that usually means energy is converted to work, energy in the form of heat moves from one place to another, or energy is stored up in the constituent chemicals. You have seen how to calculate work. Heat is defined as that energy that is transferred as a result of a temperature difference between a system and its surroundings. Mathematically, we can look at the change in energy of a system as being a function of both heat and work:

$$\Delta E = q + w$$

Where,

ΔE is the change in internal energy of a system

q is the heat flowing into the system

w is the work being done by the system

If q is positive, we say that the reaction is endothermic, that is, heat flows into the reaction from the outside surroundings. If q is negative, then the reaction is exothermic, that is, heat is given off to the external surroundings.

You might also remember the terms kinetic energy and potential energy. Kinetic energy is the energy of motion -- the amount of energy in an object that is moving. Potential energy is stationary, stored energy. If you think of a ball sitting on the edge of a table, it has potential energy in the energy possible if it falls off the table. Potential energy can be transformed into kinetic energy if and when the ball actually rolls off the table and is in motion. The total energy of the system is defined as the sum of kinetic and potential energies.

In descriptions of the energy of a system, you will also see the phrase "state properties". A state property is a quantity whose value is independent of the past history of the substance. Typical state properties are altitude, pressure, volume, temperature, and internal energy.

Enthalpy

Enthalpy is an interesting concept: it is defined by its change rather than a single entity. A state property, the word enthalpy comes from the Greek "heat inside". If you have a chemical system that undergoes some kind of change but has a fixed volume, the heat output is equal to the change in internal energy ($q = \Delta E$). We will define the enthalpy change, ΔH, of a system as being equal to its heat output at constant pressure:

$$dH = q \text{ at constant pressure}$$

Where:

$$\Delta H = \text{change in enthalpy}$$

We define enthalpy itself as,

$$H = E + PV$$

Where,

H = enthalpy

E = energy of the system

PV = pressure in atm times volume in liters

You will not need to be able to calculate the enthalpy directly; in chemistry, we are only interested in the change in enthalpy, or ΔH.

$$\Delta H = H_{final} - H_{initial} \text{ or } \Delta H = H(\text{products}) - H(\text{reactants})$$

Tables of enthalpies are generally given as ΔH values.

Example Problem: Calculate the ΔH value of the reaction

$$HCl + NH_3 \rightarrow NH_4Cl$$

(ΔH values for HCl is -92.30; NH_3 is -80.29; NH_4Cl is -314.4)

Solution

$$\Delta T = \frac{q}{MC}$$

$$\Delta T = \frac{(124)}{(2.34)(3.88)} = 13.7\,°C$$

$$\text{new } T = 22 + 13.7 = 35.7\,°C$$

We can also represent enthalpy change with the equation:

$$\Delta H = \Delta E + P\,\Delta V$$

Where,

ΔV is the change in volume, in liters

P is the constant pressure

If you recall, work is defined as P ΔV, so enthalpy changes are simply a reflection of the amount of energy change (energy going in or out, endothermic or exothermic), and the amount of work being done by the reaction. For example, if ΔE = -100 kJ in a certain combustion reaction, but 10 kJ of work needs to be done to make room for the products, the change in enthalpy is:

$$\Delta H = -100 \text{ kJ} + 10 \text{ kJ} = -90 \text{ kJ}$$

This is an exothermic reaction (which is expected with combustion), and 90 kJ of energy is released to the environment. Basically, you get warmer. Notice the convention used here -- a negative value represents energy coming out of the system.

You can also determine ΔH for a reaction based on bond dissociation energies. Breaking bonds requires energy while forming bonds releases energy. In a given equation, you must determine what kinds of bonds are broken and what kind of bonds are formed. Use this information to calculate the amount of energy used to break bonds and the amount used to form bonds. If you subtract the amount to break bonds from the amount to form bonds, you will have the ΔH for the reaction.

Example Problem: Calculate ΔH for the reaction

$$N_2 + 3H_2 \rightarrow 2NH_3$$

(The bond dissociation energy for N-N is 163 kJ/mol; H-H is 436 kJ/mol; N-H is 391 kJ/mol)

$$\Delta H = \Delta H_{products} - \Delta_{Hreactants}$$

To use the bond dissociation energies, we must determine how many bonds are in the products and the reactants. In NH_3 there are 3 N-H bonds so in 2 NH_3 there are 6 N-H bonds. In N_2 there is 1 N-N bond and in $3H_2$ there are 3 H-H bonds.

$$\Delta H_{products} = 6(391) = 2346$$
$$\Delta H_{reactants} = 163 + 3(436) = 1471$$
$$\Delta H = 2346 - 1471 = 875$$

Entropy

Entropy is a measure of the disorder of a system. Take your room as an example. Left to itself, your room will increase in entropy (i.e., get messier) if no work (cleaning up) is done to contain the disorder. Work must be done to keep the entropy of the system low. Entropy comes from the second law of thermodynamics, which states that all systems tend to reach a state of equilibrium. The significance of entropy is that when a spontaneous change occurs in a system, it will always be found that if the total entropy change for everything involved is calculated, a positive value will be obtained. Simply, all spontaneous changes in an isolated chemical system occur with an increase in entropy. Entropy, like temperature, pressure, and enthalpy, is also a state property and is represented in the literature by the symbol "S". Like enthalpy, you can calculate the change of S (ΔS).

$$\Delta S = S_{final} - S_{initial} \text{ or } \Delta S = S(\text{products}) - S(\text{reactants})$$

Where,

ΔS is change in entropy

S_{final} and $S_{initial}$ are the final and initial entropies, respectively

The following table shows the relationship between the state of a substance and its entropy:

State of substance	Relative Entropy (S)
gas	highest S
aqueous	high S
liquid	medium S
solid	lowest S

Gibbs Free Energy

Gibbs free energy is a measure of the potential for reversible or maximum work that may be done by a system at constant temperature and pressure. It is a thermodynamic property that was defined in 1876 by Josiah Willard Gibbs to predict whether a process will occur spontaneously at constant temperature and pressure. Gibbs free energy G is defined as G = H - TS where H, T, and S are the enthalpy, temperature, and entropy.

The SI unit for Gibbs energy is the kilojoule (kJ).

Changes in the Gibbs free energy G correspond to changes in free energy for processes at constant temperature and pressure. The change in Gibbs free energy change is the maximum non-expansion work obtainable under these conditions in a closed system. ΔG is negative for spontaneous processes, positive for nonspontaneous processes and zero for processes at equilibrium.

Gibbs Free Energy and Spontaneity

When a process occurs at constant temperature\text TTT and pressure \text PPP, we can rearrange the second law of thermodynamics and define a new quantity known as Gibbs free energy:

$$\text{Gibbs free energy} = G = H - TS$$

Where, H is enthalpy, T is temperature (in kelvin, K), and S is the entropy. Gibbs free energy is represented using the symbol G and typically has units of

$$\frac{\text{kJ}}{\text{mol-rxn}}$$

When using Gibbs free energy to determine the spontaneity of a process, we are only concerned with changes in G, rather than its absolute value. The change in Gibbs free energy for a process is thus written as ΔG, which is the difference between $G_{final,}$ the Gibbs free energy of the products, and $G_{initial}$ the Gibbs free energy of the reactants.

$$\Delta G = G_{final} - G_{initial}$$

For a process at constant T and constant P, we can rewrite the equation for Gibbs free energy in terms of changes in the enthalpy (ΔH_{system}) and entropy (ΔS_{system}) for our system:

$$\Delta G_{system} = \Delta H_{system} - T\Delta S_{system}$$

You might also see this reaction written without the subscripts specifying that the thermodynamic values are for the system (not the surroundings or the universe), but it is still understood that the values for ΔH and ΔS are for the system of interest. This equation is exciting because it allows us to determine the change in Gibbs free energy using the enthalpy change, ΔH, and the entropy change , ΔS, of the system. We can use the sign of ΔG to figure out whether a reaction is spontaneous in the forward direction, backward direction, or if the reaction is at equilibrium.

- When $\Delta G<0$, the process is exergonic and will proceed spontaneously in the forward direction to form more products.
- When $\Delta G>0$, the process is endergonic and not spontaneous in the forward direction. Instead, it will proceed spontaneously in the reverse direction to make more starting materials.
- When $\Delta G=0$, the system is in equilibrium and the concentrations of the products and reactants will remain constant.

Calculating change in Gibbs free energy

Although ΔG is temperature dependent, it's generally okay to assume that the ΔH and ΔS values are independent of temperature as long as the reaction does not involve a phase change. That means that if we know ΔH and ΔS we can use those values to calculate ΔG at any temperature. We won't be talking in detail about how to calculate ΔH and ΔS in this article, but there are many methods to calculate those values including:

- Estimating $\Delta H_{reaction}$ using bond enthalpies
- Calculating ΔH using standard heats of formation, $\Delta_f H^\circ$
- Calculating ΔH and ΔS using tables of standard values

When the process occurs under standard conditions (all gases at 1bar pressure, all concentrations are 1M, and T=25 °C), we can also calculate ΔG using the standard free energy of formation, $\Delta_f G^\circ$.

It is important to pay extra close attention to units when calculating ΔG from ΔH and ΔS is usually given in $\frac{kJ}{mol-reaction}$ ΔS is most often reported in $\frac{J}{mol-reaction \cdot K}$ The difference is a factor of 1000.

When is ΔG negative?

If we look at our equation in greater detail, we see that ΔG_{system} depends on 3 values:

$$\Delta G_{system}=\Delta H_{system}-T\Delta S_{system}$$

- The change in enthalpy ΔH_{system}
- The temperature T
- The change in entropy ΔS_{system}

	$\Delta H < 0$	$\Delta H > 0$
$\Delta S > 0$	Spontaneous at all T ($\Delta G < 0$)	Spontaneous at high T (when $T\Delta S$ is large)
$\Delta S < 0$	Spontaneous at low T (when $T\Delta S$ is small)	Non-spontaneous at all T ($\Delta G > 0$)

Temperature in this equation always positive (or zero) because it has units of K. Therefore, the second term in our equation, $T\Delta S_{system}$ will always have the same sign as ΔS_{system} . We can make the following conclusions about when processes will have a negative $\Delta G_{system;}$

- When the process is exothermic ($\Delta H_{system} < 0$), and the entropy of the system increases ($\Delta S_{system} > 0$), the sign of ΔG_{system} negative at all temperatures. Thus, the process is always spontaneous.
- When the process is endothermic, $\Delta H_{system} > 0$, and the entropy of the system decrease $\Delta S_{system} < 0$, the sign of ΔG is positive at all temperatures. Thus, the process is never spontaneous.

For other combinations of ΔH_{system} and ΔS_{system} , the spontaneity of a process depends on the temperature.

- Exothermic reactions ($\Delta H_{system} < 0$), that decrease the entropy of the system ($\Delta S_{system} < 0$) are spontaneous at low temperatures.
- Endothermic reactions ($\Delta H_{system} > 0$), that increase the entropy of the system ($\Delta S_{system} > 0$) are spontaneous at high temperatures.

Can you think of any reactions in your day-to-day life that are spontaneous at certain temperatures but not at others?

Example: Calculating ΔG for Melting Ice

Let's consider an example that looks at the effect of temperature on the spontaneity of a process. The enthalpy of fusion and entropy of fusion for water have the following values:

$$\Delta_{fus}H = 6.01 \frac{kJ}{mol\text{-}rxn}$$

$$\Delta_{fus}S = 22.0 \frac{J}{mol\text{-}rxn \cdot K}$$

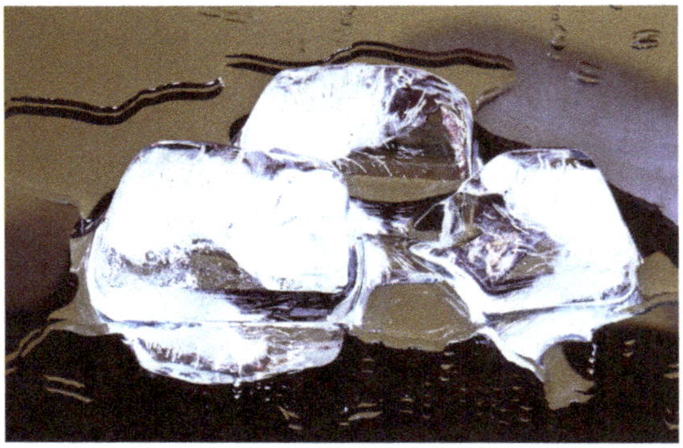

At what temperatures (if any) is the melting of ice a spontaneous process

ΔG for the Melting of Ice at 20 °C

The process we are considering is water changing phase from solid to liquid:

$$H_2O(s) \rightarrow H_2O(l)$$

For this problem, we can use the following equation to calculate ΔG_{rxn}:

$$\Delta G = \Delta H - T\Delta S$$

Luckily, we already know ΔH and ΔS for this process! We just need to check our units, which means making sure that entropy and enthalpy have the same energy units, and converting the temperature to Kelvin:

$$T = 20^{\circ}C + 273 = 293K$$

If we plug the values for ΔH,T, and ΔS into our equation, we get:

$$\Delta G = \Delta H - T\Delta S$$

$$= 6.01\frac{kJ}{mol\text{-}rxn} - (293 \times K)(0.022\frac{kJ}{mol\text{-}rxn \times \cancel{K}})$$

$$= 6.01\frac{kJ}{mol\text{-}rxn} - 6.45\frac{kJ}{mol\text{-}rxn}$$

$$= 0.44\frac{kJ}{mol\text{-}rxn}$$

Since ΔG is negative, we would predict that ice spontaneously melts at 20 °C.

Van 't Hoff Isotherm

The Gibbs free energy can change with the change of the temperature and pressure of the thermodynamic system. The Van 't Hoff isotherm can be used to determine the Gibbs free energy for non-standard state reactions at a constant temperature:

$$\left(\frac{dG}{d\varepsilon}\right)_{T,P} = \Delta_r G + RT \ln Q_r$$

where $\Delta_r G$ is the Gibbs free energy for the reaction, and Q_r is the reaction quotient. When a reaction is at equilibrium, $Q_r = K_{eq}$. The Van 't Hoff isotherm can help estimate the equilibrium reaction shift. When $\Delta_r G < 0$, the reaction moves in the forward direction. When $\Delta_r G > 0$, the reaction moves in the backwards directions.

Van 't Hoff Plot

For a reversible reaction, the equilibrium constant can be measured at a variety of temperatures. This data can be plotted on a graph with K_{eq} on the Y-axis and 1/T on the X-axis. The data should have a linear relationship, the equation for which can be found by fitting the data using the linear form of the Van 't Hoff equation

$$\ln K_{eq} = -\frac{\Delta H \Theta}{RT} + \frac{\Delta S \Theta}{R}.$$

This graph is called the Van 't Hoff plot and is widely used to estimate the enthalpy and entropy of a chemical reaction. From this plot, $-\frac{\Delta H}{R}$ is the slope and $\frac{\Delta S}{R}$ is the intercept of the linear fit.

By measuring the equilibrium constant, K_{eq}, at different temperatures, the Van 't Hoff plot can be used to assess a reaction when temperature changes. Knowing the slope and intercept from the Van 't Hoff plot, the enthalpy and entropy of a reaction can be easily obtained using,

$$\Delta H = -R \times slope,$$
$$\Delta S = R \times intercept$$

The Van 't Hoff plot can be used to quickly determine the enthalpy of a chemical reaction both qualitatively and quantitatively. Change in enthalpy can be positive or negative, leading to two major forms of the Van 't Hoff plot.

Endothermic Reactions

Endothermic Reaction Van 't Hoff Plot

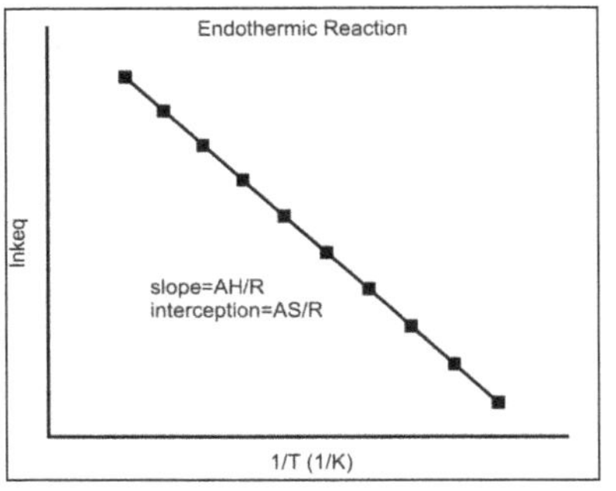

For an endothermic reaction, heat is absorbed, making the net enthalpy change positive. Thus, according to the definition of the slope:

$$Slope = -\frac{\Delta H}{R}$$

for an endothermic reaction,

$\Delta H > 0$ and R is the gas constant

So,

$$Slope = -\frac{\Delta H}{R} < 0$$

Thus, for an endothermic reaction, the Van 't Hoff plot should always have a negative slope.

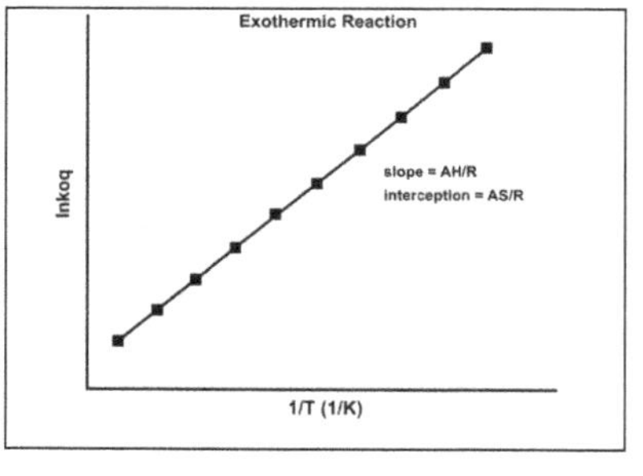

Exothermic Reaction Van 't Hoff Plot

For an exothermic reaction, heat is released, making the net enthalpy change negative. Thus, according to the definition of the slope:

$$Slope = -\frac{\Delta H}{R}$$

from an exothermic reaction,

$\Delta H < 0$ and R is the gas constant

So

$$Slope = -\frac{\Delta H}{R} > 0$$

Thus, for an exothermic reaction, the Van 't Hoff plot should always have a positive slope.

Chemical Equilibrium

A chemical reaction is in equilibrium when the concentrations of reactants and products are constant - their ratio does not vary.

Equilibrium happens when a chemical reaction does not convert all reactants to products: many reactions reach a state of balance or dynamic equilibrium in which both reactants and products are present.

Another way of defining equilibrium is to say that a system is in equilibrium when the forward and reverse reactions occur at equal rates.

Equilibrium does not necessarily mean that reactants and products are present in equal amounts. It means that the reaction has reached a point where the concentrations of the reactant and product are unchanging with time, because the forward and backward reactions have the same rate.

Equilibrium Symbol

Chemists draw arrows that look like this:

$$\rightleftharpoons$$

to show a reaction or system is in equilibrium. In the simple chemical equation below reactant A is in equilibrium with product B.

$$A \rightleftharpoons B$$

A Simple Real Example

A common example of an equilibrium is when you saturate water with table salt - in other words, you add so much salt to the water that all of the salt cannot dissolve.

Although you may think nothing much is happening in this saturated solution, at the molecular level, there is constant activity, with sodium chloride dissolving and precipitating constantly.

If the temperature is constant, equilibrium will be reached so that for every sodium chloride that dissolves, another will precipitate. At this point the system is in equilibrium, and we can write:

$$NaCl_{(solid)} \rightleftharpoons Na^{+}_{(aq)} + Cl^{-}_{(aq)}$$

The equilibrium constant indicates whether the forward reaction or backward reaction is more favored.

An equilibrium constant greater than one means the forward reaction is favored. An equilibrium constant equal to one means the forward reaction and backward reactions are equally favored. An equilibrium constant less than one means the backward reaction is favored.

Law of Chemical Equilibrium and Equilibrium Constant

Representation of the attainment of chemical equilibrium is –

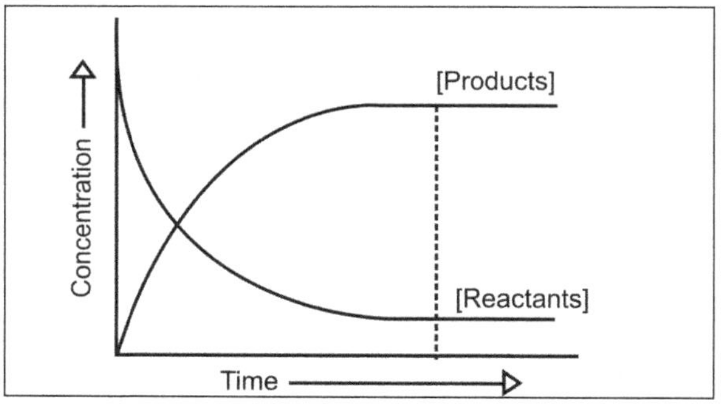

The equilibrium constant is defined as the product of the molar concentration of the products which is each raised to the power equal to its stoichiometric coefficients divided by the product of the molar concentrations of the reactants, each raised to the power equal to its stoichiometric coefficients is constant at constant temperature.

This equilibrium constant can be simply expressed in terms of the partial pressures of the reactants and the products. However, if it is expressed in terms of the partial pressure, it is denoted by K_p.

Equilibrium Constant

Equilibrium Constant Units and Formula

Law of mass action also forms the basis which states that the rate of a chemical reaction is directly proportional to the product of the concentrations of the reactants raised to their respective stoichiometric coefficients. Therefore, given the reaction –

$$aA(g) + bB(g) \Leftrightarrow cC(g) + dD(g)$$

By using the law of mass action here,

- The forward reaction rate would be $k_+ [A]^a[B]^b$
- The backward reaction rate would be $k\_ [C]^c[D]^d$

where, [A], [B], [C] and [D] being the active masses and k+ and k– are rate constants of forward and backward reactions, also the a, b, c, d are the stoichiometric coefficients related to A, B, C and D respectively. However, at the equilibrium – the forward and the backward rates are equal, stating –

Rate of forward reaction = Rate of backward reaction,

$$K_f [A]^a [B]^b = K_b [C]^c [D]^d$$

$$\frac{K_f}{K_b} = \frac{[C]^c [D]^d}{[A]^a [B]^b}$$

$$K_c = \frac{K_f}{K_b}$$

K_c is the equilibrium constant expressed in terms of the molar concentrations.

The equation $K_c = [C]^c \cdot [D]^d / [A]^a \cdot [B]^b$

or, $K_c = K_f / K_b$ is the Law of Chemical Equilibrium. The equilibrium constant is therefore related to the standard Gibbs free energy change for the reaction which is stated by the equation –

$$\S G^o = -RT \ln K_{eq}$$

Where, T states the temperature, R is the universal gas constant and K_{eq} is the equilibrium constant.

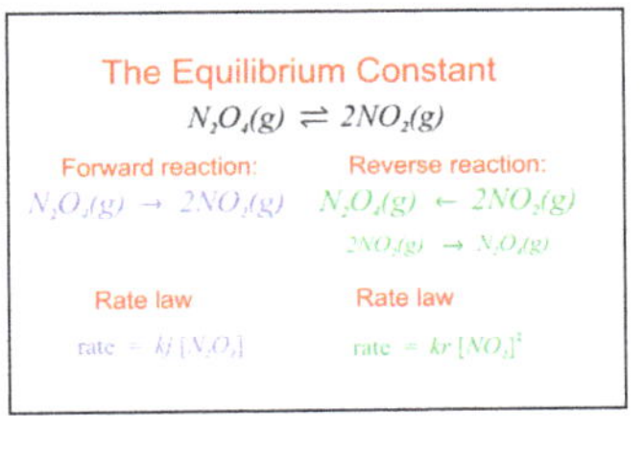

Le Chatelier's Principle

Le Chatlier's principle is also known as "Chatelier's principle" or "The Equilibrium Law". Le Chatelier's principle states that if a dynamic equilibrium is disturbed by changing the conditions (such as concentration, temperature and pressure changes) , the position of equilibrium shifts to counteract the change to reestablish an equilibrium. If a chemical reaction is at equilibrium and experiences a change in pressure, temperature, or concentration of products or reactants, the equilibrium shifts in the opposite direction to adjust itself to reduce the change.

Here we will discuss about the effects of these three changes what causes the equilibrium to shifts to minimize the stress.

1. Effect of a change in concentration
2. Effect of a change in pressure
3. Effect of change in temperature

The Effect of Concentration on Equilibrium

If the concentration of a substance is changed, the equilibrium will shift to minimise the effect of that change.

- If the concentration of a reactant is increased the equilibrium will shift in the direction of the reaction that uses the reactants, so that the reactant concentration decreases. The forward reaction is favoured.
- The forward reaction is also favoured if the concentration of the product is decreased, so that more product is formed.
- If the concentration of a reactant is decreased the equilibrium will shift in the direction of the reaction that produces the reactants, so that the reactant concentration increases. The reverse reaction is favoured.
- The reverse reaction is also favoured if the concentration of the product is increased, so that product is used.

For example, in the reaction between sulfur dioxide and oxygen to produce sulfur trioxide:

$$2SO_2(g)+O_2(g) \leftrightharpoons 2SO_3(g)$$

If the SO2 or O2 concentration was increased:

- Le Chatelier's principle predicts that equilibrium will shift to decrease the concentration of reactants.

- Increasing the rate of the forward reaction will mean a decrease in reactants.
- So some of the sulfur dioxide or oxygen is used to produce sulfur trioxide.
- Equilibrium shifts to the right. That is, when a new equilibrium is reached (when the rate of forward and reverse reactions are equal again), there will be more product than before.

If the SO_2 or O_2 concentration was decreased:

- Le Chatelier's principle predicts that the equilibrium will shift to increase the concentration of reactants.
- Increasing the rate of the reverse reaction will mean an increase in reactants.
- So some sulfur trioxide would change back to sulfur dioxide and oxygen to restore equilibrium.
- Equilibrium shifts to the left. That is, when a new equilibrium is reached there will be less product than before.

If $[SO_3]$ decreases:

- Le Chatelier's principle predicts that the equilibrium will shift to increase the concentration of products.
- Increasing the rate of the forward reaction will mean an increase in products.
- So some sulfur dioxide or oxygen is used to produce sulfur trioxide.
- Equilibrium shifts to the right. That is, when a new equilibrium is reached there will be more product than before.

If $[SO_3]$ increases:

- Le Chatelier's principle predicts that the equilibrium will shift to decrease the concentration of products.
- Increasing the rate of the reverse reaction will mean a decrease in products.
- So some of the sulfur trioxide would change back to sulfur dioxide and oxygen to restore equilibrium.
- Equilibrium shifts to the left. That is, when a new equilibrium is reached there will be less product than before.

Common-ion Effect

In solutions the change in equilibrium position can come about due to the common-ion effect. The common-ion effect is where one substance releases ions (upon dissociating or dissolving) which are already present in the equilibrium reaction.

If solid sodium chloride is added to an aqueous solution and dissolves, the following dissociation occurs:

$$NaCl(s) \rightarrow Na^{+}(aq)+Cl^{-}(aq)$$

If that solution contains the following equilibrium:

$$HCl(l) + H_2O(l) \rightleftharpoons Cl^{-}(aq)+H^{3}O^{+}(aq)$$

The added Cl^- ion (common-ion) interferes with the equilibrium by raising the concentration of the Cl^- ion. According to Le Chatelier's principle the reverse reaction speeds up as it tries to reduce the effect of the added Cl^-. As a result the equilibrium position shifts to the left.

The Effect of Temperature on Equilibrium

If the temperature of a reaction mixture is changed, the equilibrium will shift to minimise that change.

- If the temperature is increased the equilibrium will shift to favour the reaction which will reduce the temperature. The endothermic reaction is favoured.
- If the temperature is decreased the equilibrium will shift to favour the reaction which will increase the temperature. The exothermic reaction is favoured.

For example the forward reaction shown below is exothermic (shown by the negative value for ΔH). This means that the forward reaction, where nitrogen and hydrogen react to form ammonia, gives off heat, increasing the temperature (the forward reaction is exothermic). In the reverse reaction, where ammonia decomposes into hydrogen and nitrogen gas, heat is taken in by the reaction, cooling the vessel (the reverse reaction is endothermic).

$$N_2(g) + 3H_2(g) \leftrightharpoons 2NH_3(g) \quad \Delta H = -92$$

An increase in temperature:

- Favours the endothermic reaction because it takes in energy (cools the container).
- The reverse reaction is endothermic, so the reverse reaction is favoured.
- The yield of ammonia (NH_3) will decrease.

A decrease in temperature:

- Favours the exothermic reaction because it releases energy (warms the container).
- The forward reaction is exothermic, so the forward reaction is favoured.

The yield of NH_3 will increase.

The Effect of Pressure on Equilibrium

If the pressure of a gaseous reaction mixture is changed the equilibrium will shift to minimise that change.

- If the pressure is increased the equilibrium will shift to favour a decrease in pressure.
- If the pressure is decreased the equilibrium will shift to favour an increase in pressure.

When the volume of a system is decreased (and the temperature is constant), the pressure will increase. There are more collisions with the walls of the container. If there are fewer gas molecules there will be fewer collisions, and therefore lower pressure. The equilibrium will shift in a direction that reduces the number of gas molecules so that the pressure is also reduced. So, to predict in which direction the equilibrium will shift to change pressure you need to look at the number of gas molecules in the balanced reactions.

For example, the equation for the reaction between nitrogen and hydrogen is shown below:

$$N_2(g)+3H_2(g)\leftrightharpoons 2NH_3(g)$$

The ratio in the balanced equation is 1:3:2. That is, for every 1 molecule of N2 gas there are 3 molecules of H_2 gas and 2 molecules of NH_3 gas (from the balanced equation). Therefore the ratio is 4 molecules of reactant gas to 2 molecules of product gas.

An increase in pressure will:

- Favour the reaction that decreases the number of gas molecules.
- There are fewer molecules of product gas than reactant gas, so the forward reaction is favoured.
- The equilibrium will shift to the right and the yield of NH3 will increase.

A decrease in pressure will:

- Favour the reaction that increases the number of gas molecules.
- There are more molecules of reactant gas, so the reverse reaction is favoured.
- The equilibrium will shift to the left and the yield of NH3 will decrease.

Figure shows how changing the pressure of a system results in a shift in the equilibrium to counter that change. In the original system there are 12 molecules in total:

$$6H^2+2N^2\leftrightharpoons 4NH^3$$

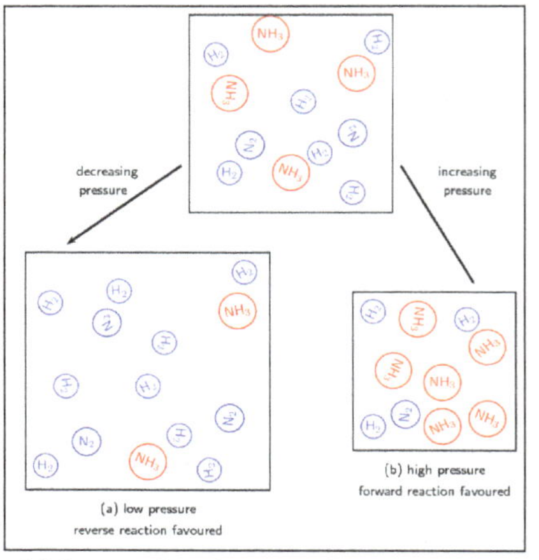

(a) A decrease in the pressure of this reaction favours the reverse reaction (more gas molecules), the equilibrium shifts to the left. (b) An increase in the pressure of this reaction favours the forward reaction (fewer gas molecules), the equilibrium shifts to the right.

If you decrease the pressure (shown by an increase in volume), the equilibrium will shift to increase the number of gas molecules. That shift is to the left and the number of H_2 and N_2 molecules will increase while the number of NH_3 molecules will decrease:

$$9H_2+3N_2 \rightleftharpoons 2NH_3 \text{ total number of molecules=14}$$

If you increase the pressure (shown by a decrease in volume), the equilibrium will shift to decrease the number of gas molecules. That shift is to the right and the number of H_2 and N_2 molecules will decrease while the number of NH_3 molecules will increase:

$$3H_2+1N_2 \rightleftharpoons 6NH_3 \text{ total number of molecules=10}$$

The total number of nitrogen and hydrogen atoms remains the same in all three situations. Equations (a) and (b) are not balanced equations.

Another example is the reaction between sulfur dioxide and oxygen:

$$2SO_2(g)+O_2(g) \leftrightharpoons 2SO_3(g)$$

In this reaction, two molecules of product gas are formed for every three molecules of reactant gas.

An increase in pressure will:

- Favour the reaction that decreases the number of gas molecules.
- There are fewer molecules of product gas than reactant gas, so the forward reaction is favoured.
- The equilibrium will shift to the right and the yield of SO_3 will increase.

A decrease in pressure will:

- Favour the reaction that increases the number of gas molecules.
- There are more molecules of reactant gas than product gas, so the reverse reaction is favoured.
- The equilibrium will shift to the left and the yield of SO3 will decrease.

The Effect of a Catalyst on Equilibrium

If a catalyst is added to a reaction, both the forward and reverse reaction rates will be increased. If both rates are increased then the concentrations of the reactants and products will remain the same. This means that a catalyst has no effect on the equilibrium position.

However, a catalyst will affect how quickly equilibrium is reached. This is very important in industry where the longer a process takes, the more money it costs. So if a catalyst reduces the amount of time it takes to form specific products, it also reduces the cost of production.

Factors that Influence the Value of Kc

Concentration, pressure, and temperature all affect the equilibrium position of a reaction, and a catalyst affects reaction rates. However, only temperature affects the value of K_c.

Changing concentration:

- Changing the concentration of a reactant or product results in one of the reactions (forward or reverse) being favoured.
- This change in reaction rate minimises the effect of the change and restores the concentration ratio between reactants and products. There will just be more reactants and products.
- K_c will remain the same.

Changing pressure:

- Changing the pressure of the system will change the ratio between the reactant and product concentrations.
- The equilibrium then shifts to minimise the effect of the change and restores the ratio between reactant and product concentrations.
- K_c will remain the same.

Adding a catalyst to the system:

- Both the forward and reverse reactions rates are increased.
- Therefore the ratio between reactant and product concentrations will remain the same.
- K_c will remain the same.

Changing temperature:

- Changing the temperature will favour either the endothermic or exothermic reaction.
- The ratio between the concentration of the reactants and products will change.

K_c will change.

References

- Thermodynamics, science: britannica.com, Retrieved 23 April, 2019
- Zeroth-law-thermodynamics: livescience.com, Retrieved 10 July, 2019
- First-law-thermodynamics: livescience.com, Retrieved 13 January, 2019
- Second_Law_of_Thermodynamics, The_Four_Laws_of_Thermodynamics, Thermodynamics, Physical_and_Theoretical_Chemistry: libretexts.org, Retrieved 11 May, 2019
- Third-law-of-thermodynamics, chemistry: byjus.com, Retrieved 10 July, 2019
- Definition-of-gibbs-free-energy: thoughtco.com, Retrieved 18 June, 2019
- Gibbs-free-energy-and-spontaneity, gibbs-free-energy, thermodynamics-chemistry, chemistry, science: khanacademy.org, Retrieved 21 March, 2019
- Van-t-hoff-isotherm: techglads.com, Retrieved 11 January, 2019
- Equilibrium: chemicool.com, Retrieved 31 March, 2019
- Law-of-chemical-equilibrium-constant, equilibrium, chemistry: toppr.com, Retrieved 15 February, 2019
- Le-chateliers-principle: chemdictionary.org, Retrieved 28 August, 2019
- Chemical-equilibrium, science: siyavula.com, Retrieved 18 March, 2019

6

Thermochemistry

Thermochemistry is a domain that studies heat energy related to chemical reactions and physical transformations. It involves the study of laws of chemistry, heat capacity, calorimetry, thermochemical equations, molar heat capacity, etc. All the diverse aspects of thermochemistry have been carefully analyzed in this chapter.

Thermochemistry is the study of the heat released or absorbed as a result of chemical reactions. It is a branch of thermodynamics and is utilized by a wide range of scientists and engineers. For example, biochemists use thermochemistry to understand bioenergetics, whereas chemical engineers apply thermochemistry to design manufacturing plants. Chemical reactions involve the conversion of a set of substances collectively referred to as "reactants" to a set of substances collectively referred to as "products." In the following balanced chemical reaction the reactants are gaseous methane CH_4 (g), and gaseous molecular oxygen, O_2 (g), and the products are gaseous carbon dioxide, CO_2 (g), and liquid water H_2O(l):

$$CH_4(g) + 2\,O_2(g) \rightarrow CO_2(g) + 2\,H_2O(1)$$

Reactions in which a fuel combines with oxygen to produce water and carbon dioxide are called combustion reactions. Because natural gas consists primarily of methane, it is expected that reaction $CH_4(g) + 2\,O_2(g) \rightarrow CO_2(g) + 2\,H_2O(1)$ will liberate heat. Reactions that liberate heat are termed exothermic reactions, and reactions that absorb heat are termed endothermic reactions.

The heat associated with a chemical reaction depends on the pressure and temperature at which the reaction is carried out. All thermochemical data presented here are for reactions carried out under standard conditions, which are a temperature of 298 °K (24.85 °C) and an applied pressure of one bar. The quantity of heat released in a reaction depends on the amount of material undergoing reaction. The chemical formulas that appear in a reaction each represent 1 mole of material; for example, the symbol CH_4 stands for 1 mole of methane having a mass of 16 grams (0.56 ounces), and the $2O_2$ (g) tells us that 2 moles of oxygen are required.

Thermochemistry also depends on the physical state of the reactants and products. For example, the heat liberated in equation,

$CH_4(g) + 2\,O_2(g) \rightarrow CO_2(g) + 2\,H_2O(1)$ is 890 kilojoules (kJ);

If, however, water in the gas phase is formed, $H_2O(g)$, the heat released is only 802 kJ. Reversing a reaction like (l), which liberates heat, yields a reaction wherein heat must be supplied for the reaction to occur. The following reaction absorbs 890 kJ.

$$CO_2(g) + 2\,H_2O(l) \rightarrow CH_4(g) + 2\,O_2(g)$$

Energy and Enthalpy

Thermochemical changes are often discussed in terms of the "system" and the "surroundings." The system is regarded as the reaction products and reactants, whereas the surroundings consist of everything else in the universe. A boundary separates the system from the surroundings. The first law of thermodynamics relates the energy change belonging to a system to the amount of work and heat crossing the boundary. A statement of the first law applied to chemical reactions in which only heat and work cross the boundary is given by the expression:

$$U_{products} - U_{reactants} = \Delta U = q + w$$

Here $U_{products}$ represents the energy of the products and $U_{reactants}$ represents the energy of the reactants. The heat associated with the reaction is given as $_q$, and w represents work done during the transformation of reactants to products. If the volume of the system changes during the reaction and the applied pressure remains constant, the work carried out is termed pressure-volume work. For example, reaction $CO_2(g) + 2\,H_2O(l) \rightarrow CH_4(g) + 2\,O_2(g)$ converts one mole of gas and two moles of liquid to a total of three moles of gas. The volume of the system increases during the reaction because, under standard conditions, a mole of gas occupies more volume than a mole of liquid. The work of expanding a system against atmospheric pressure is experienced when one inflates a balloon, and this work can be shown to be equal to $-P\,\Delta V$. Here P represents the atmospheric pressure and ΔV represents the change in volume of the system.

The first law of thermodynamics also states that U is a state function. State functions are very important in thermodynamics; they depend only on the present state of a system and not on its past history. Neither q nor w are state functions. An understanding of the concept of state function is furthered by considering the example of one's taking a trip from San Diego, California, to Denver, Colorado. The change in altitude that one experiences during this trip does not depend on the route taken and, thus, is similar to a state function. In comparison, the distance traveled between the two cities does depend on the route one follows; similarly, q and w are path-dependent quantities.

If a process such as a chemical reaction is carried out at a constant pressure in a way that involves only pressure-volume work, then $-P\Delta V$ can be substituted for the work term in equation

$$U_{products} - U_{reactants} = \Delta U = q + w$$

Thus, we have:

$$\Delta U + P \Delta V = q_p$$

The symbol q p represents the heat accompanying a chemical change carried out at constant pressure; in our previous example this would be equivalent to our specifying the exact route of travel between the two cities. The enthalpy of a system H is related to the energy of a system by the expression:

$$H = U + PV$$

For a process or reaction carried out at constant pressure:

$$\Delta H = \Delta U + P \Delta V = q_p$$

Enthalpy, like energy, is a state function. Thus, equation $\Delta H = \Delta U + P \Delta V = q_p$

shows that, for a reaction carried out at constant pressure, q p depends only on the reactants consumed and the products formed. The enthalpy change associated with a reaction carried out under standard conditions is termed the heat of reaction and is given the symbol ΔH^o, with the superscript denoting standard conditions. Endothermic reactions have a positive ΔH^o whereas exothermic reactions have a negative ΔH^o. The change in enthalpy accompanying the conversion of reactants to products in a chemical reaction determines the amount of heat liberated or absorbed by the reaction. For a reaction carried out at constant pressure the enthalpy change depends only on the reactants and products.

Hess's Law

Because enthalpy is a state function, the heat associated with a reaction does not depend on whether the reaction proceeds from reactants to products in a series of steps or in a single step. This is the basis for Hess's law, which states that if two reactions are combined to yield a third reaction, the sum of the ΔH^o s for the first two reactions is equal to the ΔH^o for the third. For example, consider the conversion of gaseous methane to liquid methanol:

$$CH_4 (g) + 1/2\, O_2 (g) \rightarrow CH_3OH\ (l)$$

and the subsequent combustion reaction:

$$CH_3OH\ (l) + 3/2\, O_2\ (g) \rightarrow 2\, H_2O(l) + CO_2\ (g)$$

Combining reactions both (above equations)by adding them together gives reaction. Thus, the ΔH^o for combined reactions

$$CH_4 (g) + 1/2\, O_2 (g) \rightarrow CH_3OH\ (l)$$

and $CH_3OH\ (l) + 3/2\, O_2\ (g) \rightarrow 2\, H_2O(l) + CO_2\ (g)$

must equal −890kJ. If the ΔH^o for reaction above is known to be −681 kJ, then the ΔH^o

for reaction $CH_4(g) + 1/2\ O_2(g) \rightarrow CH_3OH\ (l)$ can be calculated by Hess's law to equal −209 kJ. Born-Haber cycles represent an application of Hess's law to reactions associated with the formation of salts, such as potassium chloride. Born-Haber cycles can be used to determine the enthalpy change accompanying the breakup of the potassium chloride lattice into isolated potassium and chlorine ions.

Laws of Thermochemistry

The Laws of Thermodynamics and Chemistry do not apply to thermochemical systems in general. In a thermochemical system in conditions of small variations of energy and entropy the molecular variations of energy and entropy can be considered all equal for most of the molecules, however in a thermochemical system in conditions of large variations of energy and entropy the molecular variations of energy and entropy depend on the global energy and entropy, and so on the molecule.

First Law of Thermochemistry

An isolated thermochemical system evolves irreversibly to the state of thermochemical equilibrium.

$$dU \rightarrow 0 \text{ as } t \rightarrow \infty$$

in which U is the molecular energy of the isolated thermochemical system at time t.

Second Law of Thermochemistry

In a thermochemical system the energy is conserved.

$$\frac{dU}{U} = \frac{dQ + dW + dE + c^2 dm + d\gamma A}{Q + W + E + mc^2 + \gamma A}$$

in which U is the energy, Q, the thermic energy, W, the mechanical energy, E, the electromagnetic energy, mc^2, the relativistic energy and γA, the surface tension of the thermochemical system.

Third Law of Thermochemistry

In a thermochemical system the sum of the variations of the mole- cular thermo mechanical energy and the molecular kinetic energy is the sum of the variations of the molecular thermic energy and the molecular surface tension.

$$\frac{Vd + PdV + dK}{PV + K} = \frac{dQ + d\gamma A}{Q + \gamma A}$$

in which PV is the thermo mechanical energy, K, the chemical kinetic energy, Q, the thermic energy and γA, the surface tension of the thermochemical system.

Fourth Law of Thermochemistry

In a thermochemical system the molecular kinetic energy tends to half the Planck's constant times the characteristic frequency of the molecular normal mode as the molecular temperature tends to zero.

$$K \to \frac{1}{2}\hbar v \, as \; T \; \to \; 0$$

in which K is the molecular kinetic energy, $\hbar$, the Planck's constant, v, the characteristic frequency of the molecular normal mode and T, the molecular temperature of the thermochemical system.

Fifth Law of Thermochemistry

In a thermochemical system the variation of the molecular entropy is the variation of the molecular thermic energy minus the flow of molecular relativistic energy per molecular temperature.

$$\frac{dS}{S} = \frac{dQ - \Sigma a_i k \; \mu_i \; k \, d\xi_k}{(Tm / T)(Q - mc^2)}$$

in which S is the entropy, Q, the thermic energy, T, the global temperature, T_m, the molecular temperature, mc^2, the relativistic energy and a_{ik} and μ_{ik}, the stoichiometric coefficients and chemical potentials of the s chemical reactants and products in the h irreversible and reversible matter and energy exchanges and irreversible chemical reactions with extent of reaction ξ_k of the thermochemical system.

Sixth Law of Thermochemistry

In a thermochemical system the variation of the molecular entropy due to irreversible variations of the molecular thermo mechanical energy and the molecular kinetic energy and irreversible chemical reactions is always positive.

$$dS_i \geq 0$$

In which S_i is the molecular entropy of the thermochemical system due to irreversible thermochemical processes.

Seventh Law of Thermochemistry

In a thermochemical system the molecular entropy tends to zero as the molecular temperature tends to zero.

$$S \rightarrow 0 \; as \; T \rightarrow 0$$

in which S is the molecular entropy and T , the molecular temperature of the thermo-chemical system.

The van der Waals Thermochemical Equation

A differential behaviour of the van der Waals Equation on pressure in solids and liquids arises, as large variations on pressure give small variations of volume, and so, small variations of the intermolecular forces of collision, which shows that the van der Waals Thermochemical Equation for solids and liquids is

$$Pe^{aN^2/V^2}(V-nb)+\mathrm{k}=Q+\gamma A$$

in which P V is the thermo mechanical energy, K, the chemical kinetic energy, Q, the thermic energy, γA, the surface tension, N , the molecular number and a and b, the van der Waals coefficients of the thermochemical system.

The Earth Thermochemical Equations

Now we give the Earth Thermochemical Equations.

By the Third Law of Thermochemistry, since the atmosphere is a thermochemical system with no varying heat capacity within a spherical ring volume with a radius H for which small variations of temperature occur within large variations of pressure, small variations of radius and small variations of chemical kinetic energy, its thermochemical differential equation is,

$$\frac{H_o dP + PdH + dK}{HoP + K_o} = \frac{dT}{To}$$

and so,

$$\left(H_0 P \; + \; K_0\right) e^{(HP \; +K)=(H_0 P \; +K_0)} = \left(H_0 P_0 + K_0\right) e^{\,T/T0}$$

in which K is the chemical kinetic energy, P , the pressure, T , the temperature, and H, the radius of the atmosphere, and K_o, the chemical kinetic energy, P_o, the pressure, T_o, the temperature and H_o, the radius of the atmosphere at sea level.

By the Third Law of Thermochemistry, since the Earth is a thermochemical system with no varying heat capacity in which small variations of temperature and no variations of surface tension occur within large variations of pressure, large variations of radius and small variations of chemical kinetic energy, its thermochemical differential equation is,

$$\frac{HdP + PdH + dk}{HP + K_0} = \frac{dT}{T_0 + \gamma A_0}$$

and so,

$$(HP+K_0)e^{K/(HP+K_O)+T_O(T_O+\gamma A_0)}(H_0P_0+K_0)e^{T/(T_0+\gamma A_0)+K_0(HP+K_0)}$$

in which K is the chemical kinetic energy, P, the pressure, T, the temperature and H, the radius of the Earth, and K_0, the chemical kinetic energy, P_0, the pressure, T_0, the temperature, H_0, the radius and ãA_0, the surface tension of the Earth at sea level.

By the Third Law of Thermochemistry, since the nucleus of Earth is a thermochemical system with varying heat capacity in which large variations of thermic energy and surface tension occur within large variations of pressure, small variations of radius and large variations of chemical kinetic energy, its thermochemical differential equation is

$$\frac{\frac{4}{3}\pi H_0^3\,dP+4\pi H^2PdH+dK}{\frac{4}{3}\pi H_0^3\,P+K}=\frac{dQ+d\gamma A}{Q+\gamma A}$$

and so,

$$(Q_0+A_0)\left(\frac{4}{3}\pi H_0^3P+K\right)e^{\frac{4}{3}\pi H^3P/\left(\frac{4}{3}\pi H_0^3P+K\right)}=\left(\frac{4}{3}\pi H_0^3P_o+K_o\right)(Q+\gamma A)e^{\frac{4}{3}\pi H_0^3P/\left(\frac{4}{3}\pi H_0^3P+K\right)}$$

in which K is the chemical kinetic energy, Q, the thermic energy, P , the pressure, H, the radius and , the surface tension of the magmatic core and K_0, the chemical kinetic energy, Q_0, the thermic energy, P_0, the pressure, H_0, the radius and γA_0, the surface tension of the magmatic core surface.

Quantum Thermochemistry

In Quantum Thermochemistry, the relativistic energy of a quantum thermochemical system is the thermo mechanical energy plus the chemical kinetic energy,

$$mc^2=PV+\frac{1}{2}\hbar v+\frac{\hbar v}{e^{\hbar v}/kT-1}$$

in which mc^2 is the relativistic energy, PV , the thermo mechanical energy, k, the Planck's constant, v, the chacteristic frequency of the normal mode, $\hbar$, the Boltzmann's constant and T , the temperature of the quantum thermochemical system.

Calorimetry

One technique we can use to measure the amount of heat involved in a chemical or physical process is known as calorimetry. Calorimetry is used to measure amounts of heat transferred to or from a substance. To do so, the heat is exchanged with a calibrated object (calorimeter). The change in temperature of the measuring part of the calorimeter is converted into the amount of heat (since the previous calibration was used to establish its heat capacity). The measurement of heat transfer using this approach requires the definition of a system (the substance or substances undergoing the chemical or physical change) and its surroundings (the other components of the measurement apparatus that serve to either provide heat to the system or absorb heat from the system). Knowledge of the heat capacity of the surroundings, and careful measurements of the masses of the system and surroundings and their temperatures before and after the process allows one to calculate the heat transferred.

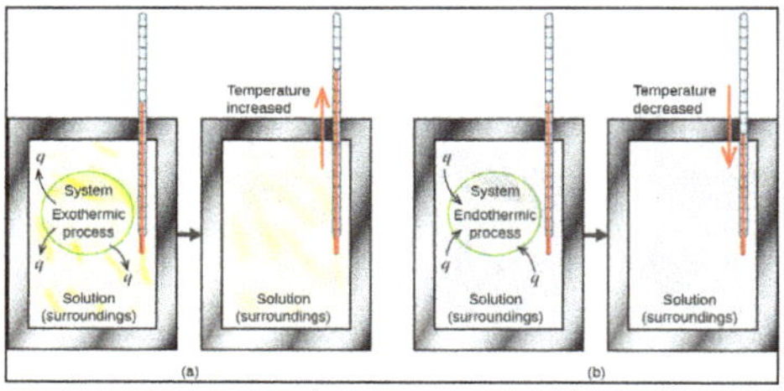

In a calorimetric determination, either (a) an exothermic process occurs and heat, q, is negative, indicating that thermal energy is transferred from the system to its surroundings, or (b) an endothermic process occurs and heat, q, is positive, indicating that thermal energy is transferred from the surroundings to the system.

A calorimeter is a device used to measure the amount of heat involved in a chemical or physical process. For example, when an exothermic reaction occurs in solution in a calorimeter, the heat produced by the reaction is absorbed by the solution, which increases its temperature. When an endothermic reaction occurs, the heat required is absorbed from the thermal energy of the solution, which decreases its temperature (Figure). The temperature change, along with the specific heat and mass of the solution, can then be used to calculate the amount of heat involved in either case.

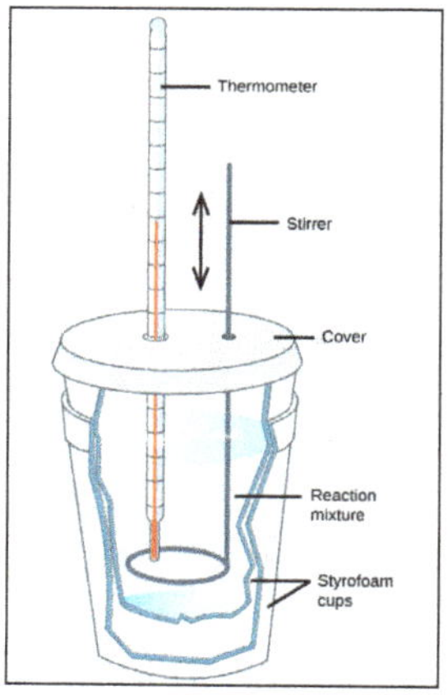

Figure: A simple calorimeter can be constructed from two polystyrene cups. A thermometer and stirrer extend through the cover into the reaction mixture.

Scientists use well-insulated calorimeters that all but prevent the transfer of heat between the calorimeter and its environment. This enables the accurate determination of the heat involved in chemical processes, the energy content of foods, and so on. General chemistry students often use simple calorimeters constructed from polystyrene cups. These easy-to-use "coffee cup" calorimeters allow more heat exchange with their surroundings, and therefore produce less accurate energy values.

Commercial solution calorimeters are also available. Relatively inexpensive calorimeters often consist of two thin-walled cups that are nested in a way that minimizes thermal contact during use, along with an insulated cover, handheld stirrer, and simple thermometer. More expensive calorimeters used for industry and research typically have a well-insulated, fully enclosed reaction vessel, motorized stirring mechanism, and a more accurate temperature sensor.

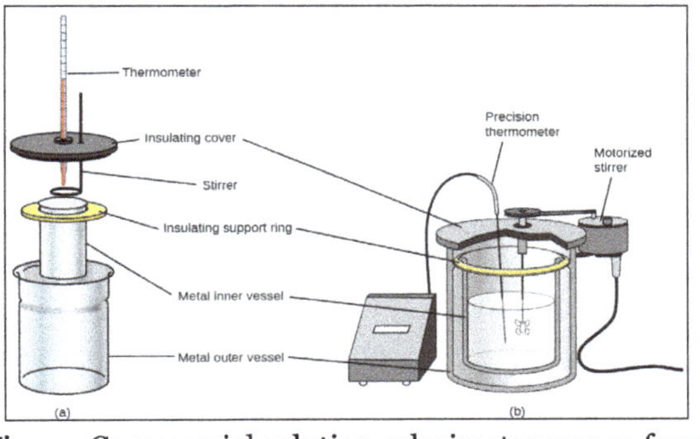

Figure: Commercial solution calorimeters range from
(a) simple, inexpensive models for student use to
(b) expensive, more accurate models for industry and research.

Before we practice calorimetry problems involving chemical reactions, consider a simpler example that illustrates the core idea behind calorimetry. Suppose we initially have a high-temperature substance, such as a hot piece of metal (M), and a low-temperature substance, such as cool water (W). If we place the metal in the water, heat will flow from M to W. The temperature of M will decrease, and the temperature of W will increase, until the two substances have the same temperature—that is, when they reach thermal equilibrium. If this occurs in a calorimeter, ideally all of this heat transfer occurs between the two substances, with no heat gained or lost by either the calorimeter or the calorimeter's surroundings. Under these ideal circumstances, the net heat change is zero:

$$q_{\text{substance M}} + q_{\text{substance W}} = 0$$

This relationship can be rearranged to show that the heat gained by substance M is equal to the heat lost by substance W:

$$q_{\text{substance M}} = -q_{\text{substance W}} = 0$$

The magnitude of the heat (change) is therefore the same for both substances, and the negative sign merely shows that $q_{\text{substance M}}$ and $q_{\text{substance W}}$ are opposite in direction of heat flow (gain or loss) but does not indicate the arithmetic sign of either q value (that is determined by whether the matter in question gains or loses heat, per definition). In the specific situation described, $q_{\text{substance M}}$ is a negative value and $q_{\text{substance W}}$ is positive, since heat is transferred from M to W.

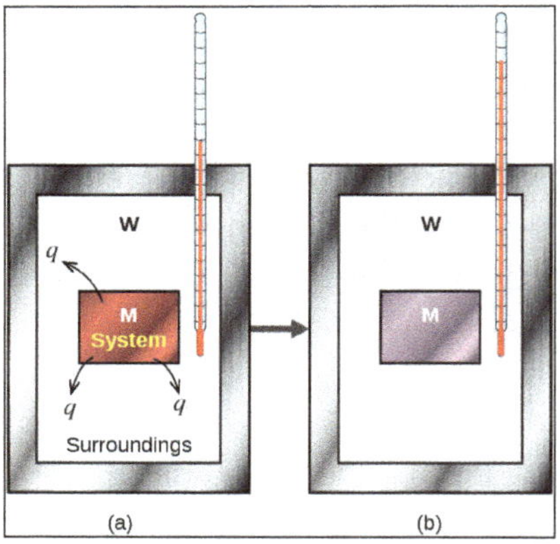

Figure. In a simple calorimetry process, (a) heat, q, is transferred from the hot metal, M, to the cool water, W, until (b) both are at the same temperature.

Calorimeter Principle

When two bodies of different temperatures (preferably a solid and a liquid) are placed in physical contact with each other, the heat is transferred from the body with higher temperature to the body with lower temperature until thermal equilibrium is attained between them. The body at higher temperature releases heat while the body at lower temperature absorbs heat. The principle of calorimetry indicates the law of conservation energy, i.e. the total heat lost by the hot body is equal to the total heat gained by the cold body.

Heat Lost = Heat Gained

The heat transfer in a system is calculated using the formula,

$$q = mc\Delta t$$

Where,

q is the is the measure of heat transfer;

m is the mass of the body;

c is the specific heat of the body;

Δt is the change in the temperature.

Calorimeter Problems

Let us look at the example to understand how to calculate the heat transfer between two objects.

Example: A metal weighing 4.82 g was heated to 115.0 °C and put into 35 mL of water of temperature 28.7 °C. The metal and water were allowed to come to an equilibrium temperature, determined to be 34.5 °C. Assuming no heat was lost to the environment, calculate the specific heat of the metal. Consider the specific heat capacity of water as 4.186 joule/gram °C.

Solution: First, let us calculate the heat absorbed by the water and use the value obtained to calculate the specific heat of the metal.

The formula to find the heat absorbed by the water is given as

$$q = mc\Delta t$$

Substituting the values in the equation, we get

$$q_{absorbed} = (4.186J / g \cdot {}^{\circ}C) \times 35g \times (5.8\,{}^{\circ}C)$$
$$q_{absorbed} = 850J = q_{released}$$

Now, using this formula let us calculate the specific heat of the metal as follows:

$$850J = s \times 4.82g \times 80.5\,{}^{\circ}C$$
$$s = 2.19J / G \cdot {}^{\circ}C$$

The measurements obtained using the principle of calorimetry explain a lot of important phenomenon in thermodynamics

Heat Capacity

Heat capacity is the ratio of heat absorbed by a material to the temperature change. It is usually expressed as calories per degree in terms of the actual amount of material being considered, most commonly a mole (the molecular weight in grams). The heat capacity in calories per gram is called specific heat. The definition of the calorie is based on the specific heat of water, defined as one calorie per degree Celsius.

At sufficiently high temperatures, the heat capacity per atom tends to be the same for all elements. For metals of higher atomic weight, this approximation is already a good one at room temperature, giving rise to the law of Dulong and Petit. For other materials, heat capacity and its temperature variation depend on differences in energy levels for atoms (available quantum states). Heat capacities are measured with some variety of calorimeter, and, using the formulation of the third law of thermodynamics; heat-capacity measurements became important as a means of determining the entropies of various materials.

Specific Heat Capacity

Specific heat capacity is the amount of heat energy required to raise the temperature of a substance per unit of mass. The specific heat capacity of a material is a physical property. It is also an example of an extensive property since its value is proportional to the size of the system being examined.

In SI units, specific heat capacity (symbol: c) is the amount of heat in joules required to raise 1 gram of a substance 1 Kelvin. It may also be expressed as J/kg·K. Specific heat capacity may be reported in the units of calories per gram degree Celsius, too. Related values are molar heat capacity, expressed in J/mol·K, and volumetric heat capacity, given in $J/m^3 \cdot K$.

Heat capacity is defined as the ratio of the amount of energy transferred to a material and the change in temperature that is produced:

$$C = Q / \Delta T$$

Where C is heat capacity, Q is energy (usually expressed in joules), and ΔT is the change in temperature (usually in degrees Celsius or in Kelvin). Alternatively, the equation may be written:

$$Q = Cm\,\Delta T$$

Specific heat and heat capacity are related by mass:

$$C = m \times S$$

Where C is heat capacity, m is mass of a material, and S is specific heat. Note that since specific heat is per unit mass, its value does not change, no matter the size of the sample. So, the specific heat of a gallon of water is the same as the specific heat of a drop of water.

It's important to note the relationship between added heat, specific heat, mass, and temperature change does not apply during a phase change. The reason for this is because heat that is added or removed in a phase change does not alter the temperature.

Specific Heat Capacity Examples

Water has a specific heat capacity of 4.18 J (or 1 calorie/gram °C). This is a much higher value than that of most other substances, which makes water exceptionally good at regulating temperature. In contrast, copper has a specific heat capacity of 0.39 J.

Table of Common Specific Heats and Heat Capacities.

This chart of specific heat and heat capacity values should help you get a better sense of the types of materials that readily conduct heat versus those which do not. As you might expect, metals have relatively low specific heats.

Material	Specific Heat (J/g°C)	Heat Capacity (J/°C for 100 g)
gold	0.129	12.9
mercury	0.140	14.0
copper	0.385	38.5
iron	0.450	45.0
salt (Nacl)	0.864	86.4
aluminum	0.902	90.2
air	1.01	101
ice	2.03	203
water	4.179	417.9

Molar Heat Capacity

Molar heat capacity or molar specific heat capacity is the amount of heat energy required to raise the temperature of 1 mole of a substance.

In SI units, molar heat capacity (symbol: c_n) is the amount of heat in joules required to raise 1 mole of a substance 1 Kelvin.

$$c_n = Q / \Delta T$$

Where Q is heat and ΔT is the change in temperature. For most purposes, heat capacity is reported as an intrinsic property, meaning it is a characteristic of a specific substance. Heat capacity is measured using a calorimeter. A bomb calorimeter is used for calculations at constant volume. Coffee cup calorimeters are appropriate for finding constant pressure heat capacity.

Units of Molar Heat Capacity

Molar heat capacity is expressed in units of J/K/mol or J/mol·K, where J is joules, K is Kelvin, and m is number of moles. The value assumes no phase changes occur. You'll typically start out with the value for molar mass, which is in units of kg/mol. A less common unit of heat is the kilogram-Calorie (Cal) or the cgs variant, the gram-calorie (cal). It's also possible to express heat capacity in terms of pound-mass using temperatures in degrees Rankine or Fahrenheit.

Molar Heat Capacity Examples

Water has a molar specific heat capacity of 75.32 J/mol·K. Copper has a molar specific heat capacity of 24.78 J/mol·K.

Molar Heat Capacity Versus Specific Heat Capacity

While molar heat capacity reflects the heat capacity per mole, the related term specific heat capacity is the heat capacity per unit mass. Specific heat capacity is also known simply as specific heat. Sometimes engineering calculations apply volumetric heat capacity, rather than specific heat based on mass.

Enthalpy Change

Enthalpy change is the heat change accompanying a chemical reaction at constant volume or constant pressure. The enthalpy change tells the amount of heat absorbed or evolved during the reaction. It is denoted by ΔH.

Types of Enthalpy Change

Enthalpy change of a reaction expressed in different ways depending on the nature of the reaction.

Heat of Formation

The enthalpy change that takes place when one mole of compound is formed from its elements. It is denoted by ΔHf. As for example, the heat of formation of ferrous sulphide from its elements is given:

$$Fe_{(s)} + S_{(s)} \rightarrow FeS_{(s)} \ \Delta H_f = -24.0 \text{ kcal}$$

Heat of combustion

The enthalpy change that takes place when one mole of compound is completely burnt in excess of air or oxygen. It is denoted by ΔHc. As for example, the heat of combustion of methane is given :

$$CH_{4(g)} + 2O_{2(g)} \rightarrow CO_{2(g)} + 2H_2O_{(l)} \ \Delta H_c = -21.0 \text{ kcal}$$

Heat of Neutralization

The enthalpy change that takes place when one gram equivalent of an acid is completely neutralized with one gram equivalent of base in dilute solution. As for example, the heat of neutralization of nitric acid and sodium hydroxide is given:

$$HNO_{3(aq)} + NaOH_{(aq)} \rightarrow NaNO_{3(aq)} + H_2O_{(l)} \ \Delta H = -13.69 \text{ kcal}$$

Heat of Solution

The enthalpy change that takes place when one mole of substance is dissolved in specified quantity of solvent in given temperature. As for example, the heat of solution of magnesium sulphate is given:

$$MgSO_{4(s)} + H_2O_{(l)} \rightarrow MgSO_{4(aq)} \ \Delta H = -20.28 \text{ kcal}$$

Calculating Enthalpy Change

The enthalpy of a system is measured by the sum of the internal energy and the product at constant pressure and volume.

$$H = E + PV$$

Here E = internal energy, P = Pressure and V = volume of the system. It is also called heat content. Thus the enthalpy change is

$$\Delta H = H_{products} - H_{reactants}$$

$$\Delta H = H_p - H_r$$

If ΔV = the change in volume at constant temperature and pressure, ΔE = the sum of the change in internal energy at constant volume and ΔH = the change in enthalpy at constant pressure. Then from the equation (H = E + PV) we can write,

$$\Delta H = \Delta E + P\Delta V$$

Let us consider a reaction,

$$aA + bB \rightarrow cC + dD$$

Here,

Change in number of moles = no.of moles of products – no. of moles of reactants.

$$= (c + d) - (a + b)$$

$$= \Delta n$$

If V = the volume occupied by one mole of gas, the change in volume,

ΔV = change in no. of moles X volume occupied by one mole of gas.

$$\Delta V = \Delta n \text{ X } V$$

$$P\Delta V = P(\Delta n \times V)$$

$$P\Delta V = PV \times \Delta n$$

But for one mole of gas PV = RT. By placing this value in equation ($P\Delta V = PV \times \Delta n$) we get

$$P\Delta V = RT \times \Delta n$$

Substituting the value of $P\Delta V$ in equation ($\Delta H = \Delta E + P\Delta V$) we get

$$\Delta H = \Delta E + \Delta n\ RT$$

To calculate Δn, the no. of moles of reactants and products in only gaseous state are considered. The value of gas constant

$$R = 8.314\ J/mol.K \text{ or } 1.987\ cal/mol.K.$$

Exothermic and Endothermic Reactions

Let us consider a reaction at constant pressure,

$$A + B \rightarrow C + D$$

Rewritting equation ($\Delta H = Hp - Hr$) we get,

$$\Delta H = Hp - Hr$$

$$\Delta H = (H_C + H_D) - (H_A + H_B)$$

The value of ΔH can be positive, negative or zero. If ΔH is zero that means there is no heat change during the reaction, neither heat is absorbed or evolved.

Sign of ΔH	Energy	Reaction type
+	Absorbed	Endothermic
–	Evolved	Exothermic

In case ΔH is positive that means Hp is greater than Hr, then heat is absorbed or the reaction is endothermic. And when ΔH is negative means Hp is less than Hr then heat is evolved or the reaction is exothermic.

Enthalpy of Formation and Combustion

Enthalpy of Formation

Enthalpy of formation (ΔHf) mentions the quantity of energy essential to produce a compound from its composition of elements. The standard enthalpy of formation (ΔH_f^o) refers to the quantity of energy essential to produce one mole of a mixture from its composition of elements.

Example

One half mole of gaseous hydrogen $H_2(g)$ reacts with one half mole of gaseous chlorine $Cl_2(g)$ to produce one mole of gaseous hydrogen chloride HCl(g) with the enthalpy of formation of – 92.30 kJ.

$$\Delta H_f = -92.30\text{kJ}$$

Show the creation of carbon dioxide from methane with the enthalpy of formation values as in figure.

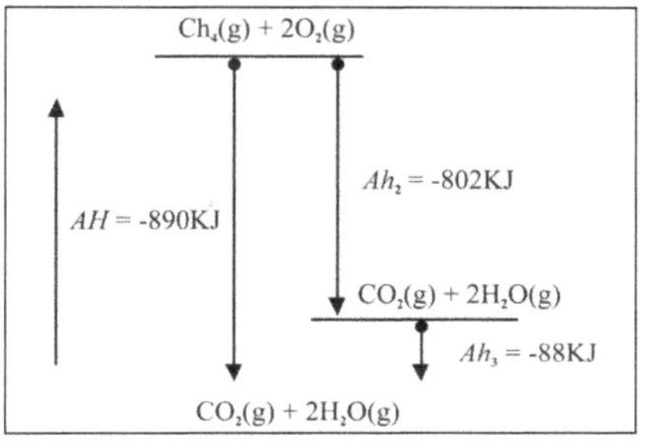

Enthalpy of Combustion

Enthalpy of combustion denotes the change in enthalpy experienced when one mole of an element is heated in excess amount of oxygen under the standard conditions. It is related to the combustion of the element throughout the process. It is symbolized by Δ Hc.

Example

When magnesium in solid state in heated in oxygen, then magnesium oxide is created with the negative value of enthalpy of combustion.

$$\text{Mg (s)} + \frac{1}{2}\text{o}_2(g) \rightarrow \text{Mgo } \Delta H_c = -\text{ve}$$

Standard Enthalpy Changes: Standard Conditions

To compare the enthalpy changes of a various reactions we must use standard conditions, such as known temperatures, pressures, amounts and concentrations of reactants or products.

The standard conditions are:

- A pressure of 100 kilopascals (10^2kPa);

- A temperature of 298K (25 °C);
- Reactants and products in physical states, normal for the above conditions;
- A concentration of 1.0mol dm^{-3} for solutions.

The ° sign indicates standard conditions.

Standard Enthalpy Change of Reaction

$$\Delta H_r^o$$

The standard enthalpy change of reaction is the enthalpy change when the amounts of reactants shown in the equation for the reaction, react under standard conditions to give the products in their standard states.

Standard Enthalpy Change of Formation

$$\Delta H_f^o$$

The standard enthalpy change of formation is the enthalpy change when one mole of a compound is formed from its elements under standard conditions; both compound and elements are in standard states.

Standard Enthalpy Change of Combustion

$$\Delta H_f^o$$

The standard enthalpy change of combustion is the enthalpy change when one mole of an element or compound reacts completely with oxygen under standard conditions.

Thermochemical Equations

Energy changes which accompany chemical reactions are almost always expressed by thermochemical equations. When methane gas is combusted, heat is released, making the reaction exothermic. Specifically, the combustion of 1 mol of methane releases 890.4 kilojoules of heat energy. This information can be shown as part of the balanced equation.

$$CH_4(g)+2O_2(g)\rightarrow CO_2(g)+2H_2O(l)+890.4kJ$$

The equation tells us that 1 mol of methane combines with 2 mol of oxygen to produce 1 mol of carbon dioxide and 2 mol of water. In the process, 890.4 kJ is released and so it is written as a product of the reaction. A thermochemical equation is a chemical equation that includes the enthalpy change of the reaction. The process in the above thermochemical equation can be shown visually in the figure below.

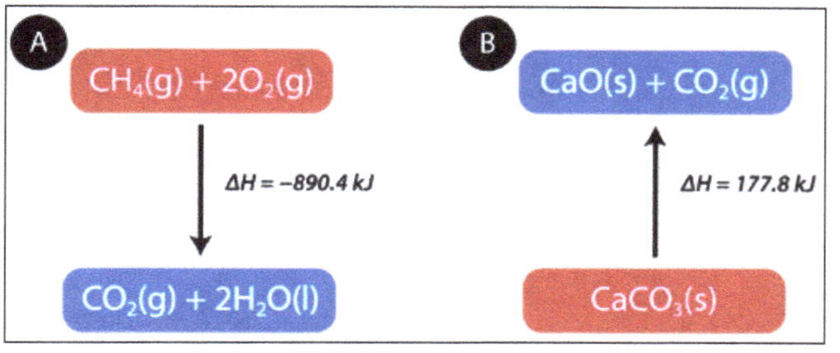

(A) As reactants are converted to products in an exothermic reaction, enthalpy is released into the surroundings. The enthalpy change of the reaction is negative.

(B) As reactants are converted to products in an endothermic reaction, enthalpy is absorbed from the surroundings. The enthalpy change of the reaction is positive.

In the combustion of methane example, the enthalpy change is negative because heat is being released by the system. Therefore, the overall enthalpy of the system decreases. The heat of reaction is the enthalpy change for a chemical reaction. In the case above, the heat of reaction is −890.4 kJ. The thermochemical reaction can also be written in this way:

$$\mathrm{CH_4}(g)+2\mathrm{O_2}(g)\rightarrow\mathrm{CO_2}(g)+2\mathrm{H_2O}(l)\ \Delta H=-890.4\mathrm{kJ}$$

Heats of reaction are typically measured in kilojoules. It is important to include the physical states of the reactants and products in a thermochemical equation as the value of the ΔH depends on those states.

Endothermic reactions absorb energy from the surroundings as the reaction occurs. When 1 mol of calcium carbonate decomposes into 1 mol of calcium oxide and 1 mol of carbon dioxide, 177.8 kJ of heat is absorbed. The process is shown visually in figure above. The thermochemical reaction is shown below,

$$\mathrm{CaCO_3(s)}+177.8\mathrm{kJ}\rightarrow\mathrm{CaO(s)}+\mathrm{CO_2}(g)$$

Because the heat is absorbed by the system, the 177.8 kJ is written as a reactant. The heat of reaction is positive for an endothermic reaction.

$$CaCO_3(s)\rightarrow CaO(s)+CO_2(g)\quad \Delta H=177.8\,kJ$$

The way in which a reaction is written influences the value of the enthalpy change for the reaction. Many reactions are reversible, meaning that the product(s) of the reaction are capable of combining and reforming the reactant(s). If a reaction is written in the reverse direction, the sign of the ΔH changes. For example, we can write an equation for the reaction of calcium oxide with carbon dioxide to form calcium carbonate.

$$CaO(s) + CO_2(g) \rightarrow CaCO_3(s) + 177.8kJ$$

The reaction is exothermic and thus the sign of the enthalpy change is negative.

$$CaO(s) + CO_2(g) \rightarrow CaCO_3(s) \Delta H = -177.8kJ$$

References

- Thermochemistry: chemistryexplained.com, Retrieved 14 February, 2019
- Calorimetry, chemistry: opentextbc.ca, Retrieved 6 July, 2019
- Principle-of-calorimetry, physics: byjus.com, Retrieved 21 May, 2019
- Heat-capacity, science: britannica.com, Retrieved 8 March, 2019
- Definition-of-specific-heat-capacity: thoughtco.com, Retrieved 12 June, 2019
- Definition-of-molar-heat-capacity-and-examples: thoughtco.com, Retrieved 15 August, 2019
- Enthalpy-change: chemdictionary.org, Retrieved 29 January, 2019
- Enthalpy-of-formation, enthalpy-of-combustion-and-the-first-law: chegg.com, Retrieved 3 April, 2019
- Enthalpy-changes, chemical-energetics, chemistry: s-cool.co.uk, Retrieved 9 March, 2019
- Thermochemical-equation: lumenlearning.com, Retrieved 13 August, 2019

7

Electrochemistry

Electrochemistry deals with the interaction between electrical energy and chemical change. Electrochemical cell is a tool via which electric current is produced from energy that is released by a spontaneous redox reaction. This chapter discusses in detail the theories related to electrochemistry.

Electrochemistry is the study of chemical processes that cause electrons to move. This movement of electrons is called electricity, which can be generated by movements of electrons from one element to another in a reaction known as an oxidation-reduction ("redox") reaction. A redox reaction is a reaction that involves a change in oxidation state of one or more elements. When a substance loses an electron, its oxidation state increases; thus, it is oxidized.

When a substance gains an electron, its oxidation state decreases, thus being reduced. For example, for the redox reaction,

$$H_2 + F_2 \rightarrow 2HF$$

can be rewritten as follows:

- Oxidation reaction

$$H_2 \rightarrow 2H^+ + 2e^-$$

- Reduction reaction:

$$F_2 + 2e^- \rightarrow 2F^-$$

- Overall Reaction

$$H_2 + F_2 \rightarrow 2H^+ + 2F^-$$

Oxidation is the loss of electrons, whereas reduction refers to the acquisition of electrons, as illustrated in the respective reactions above. The species being oxidized is also known as the reducing agent or reductant, and the species being reduced is called the oxidizing agent or oxidant. In this case, H2 is being oxidized (and is the reducing agent), while F_2 is being reduced (and is the oxidizing agent).

Electrochemical Cells

An electrochemical cell is a device that produces an electric current from energy released by a spontaneous redox reaction. This kind of cell includes the galvanic, or voltaic, cell, named after Luigi Galvani and Alessandro Volta. These scientists conducted several experiments on chemical reactions and electric current during the late 18th century.

Electrochemical cells have two conductive electrodes, called the anode and the cathode. The anode is defined as the electrode where oxidation occurs. The cathode is the electrode where reduction takes place. Electrodes can be made from any sufficiently conductive materials, such as metals, semiconductors, graphite, and even conductive polymers. In between these electrodes is the electrolyte, which contains ions that can freely move.

The voltaic cell uses two different metal electrodes, each in an electrolyte solution. The anode will undergo oxidation and the cathode will undergo reduction. The metal of the anode will oxidize, going from an oxidation state of 0 (in the solid form) to a positive oxidation state, and it will become an ion. At the cathode, the metal ion in the solution will accept one or more electrons from the cathode, and the ion's oxidation state will reduce to 0. This forms a solid metal that deposits on the cathode. The two electrodes must be electrically connected to each other, allowing for a flow of electrons that leave the metal of the anode and flow through this connection to the ions at the surface of the cathode. This flow of electrons is an electrical current that can be used to do work, such as turn a motor or power a light.

Two Types of Cell

There are two fundamental types of electrochemical cell: galvanic and electrolytic.

- Galvanic cells convert chemical potential energy into electrical energy. The energy conversion is achieved by spontaneous ($\Delta G < 0$) redox reactions producing a flow of electrons.
- Electrolytic cells are driven by an external source of electrical energy. A flow of electrons drives non-spontaneous ($\Delta G \geq 0$) redox reactions.

Galvanic Cells

A galvanic cell is a cell where chemical reactions between dissimilar conductors connected through an electrolyte and a salt bridge produce electric energy. A galvanic cell can also be powered by spontaneous oxidation-reduction reactions. Essentially, a galvanic cell channels the electrical energy produced by the electron transfer in a redox reaction. The electrical energy or current may be sent to a circuit, such as in a television or light bulb.

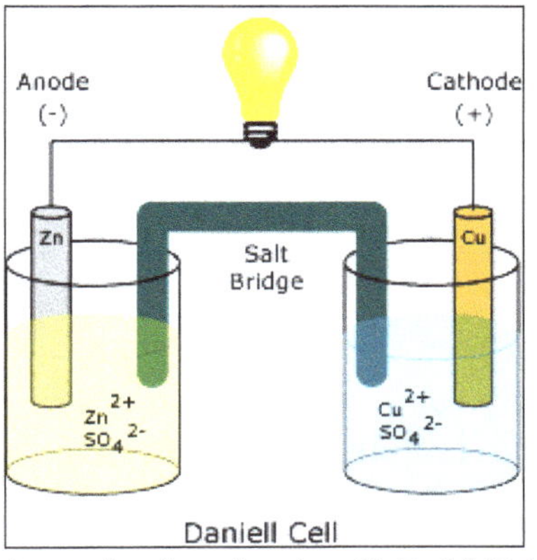

General schematic of a galvanic cell. Todd Helmenstine

The electrode of the oxidation half-cell is the anode (-), while the electrode of the reduction half-cell is the cathode (+). The mnemonic "The Red Cat Ate an Ox" may be used to help remember reduction occurs at the cathode and oxidation occurs at the anode.

A galvanic cell is also called a Daniel cell or a voltaic cell.

Electrolytic Cells

Electrolytic cells convert electrical energy into chemical potential energy. The process is known as electrolysis. The purpose of this is usually to convert reactants into more useful products.

Electrolytic cells are one of two major categories of electrochemical cell.

The other category, voltaic cells, convert chemical potential energy to electrical energy. Battery-powered devices get their electrical energy from one or more voltaic cells.

Thermodynamics

Electrolytic cells use electrical energy to drive non-spontaneous chemical reactions - i.e. those reactions for which $\Delta G \geq 0$.

Examples of Electrolytic Cells

Manufacturing Sodium

Electrolysis is used to manufacture sodium metal from sodium chloride

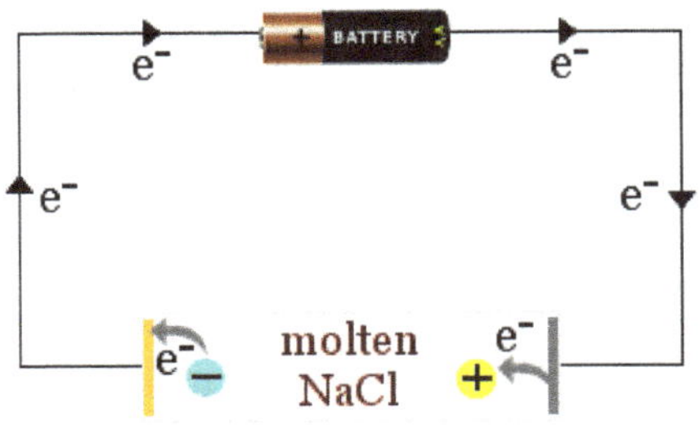

$$Cl^-_{(l)} \rightarrow \frac{1}{2}Cl_{2(g)} + e^- \quad Na^+_{(l)} + e^- \rightarrow Na_{(l)}$$

Recharging Batteries

Electrolysis is used to recharge rechargeable batteries: rechargeable batteries operate as voltaic cells when they are powering devices and as electrolytic cells during recharge.

For example, the Edison battery is a simple, rechargeable cell invented by Thomas Edison. It consists of two metal electrodes, one made of iron, the other of nickel. During initial charging, a coating of nickel oxide forms on the nickel electrode.

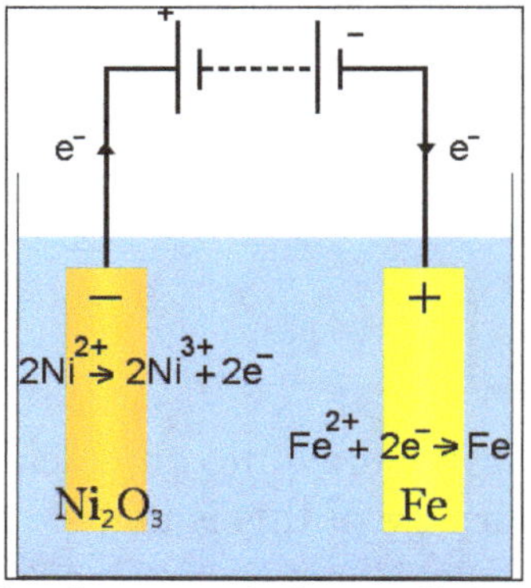

Crude schematic of Edison cell charging

The electrolyte (the ionic liquid between the electrodes) is aqueous potassium hydroxide.

When it is discharging, the Edison cell operates as a voltaic cell. When it is being charged, the cell operates as an electrolytic cell.

The chemical equations for the reactions at the electrodes are:

$$Ni_2O_3 + H_2O + 2\,e^- \rightleftharpoons 2\,NiO + 2\,OH^-$$

$$Fe + 2\,OH^- \rightleftharpoons Fe(OH)_2 + 2\,e^-$$

During discharge, when the cell is delivering electrical energy, the reactions above proceed from left to right.

During charging, when the cell is operating electrolytically, converting electrical energy to chemical potential energy, the reactions above proceed from right to left.

Electrolysis of Water

Electrolysis of water produces the ultimate in clean fuels; one of chemistry's Holy Grails is to split water using sunlight.

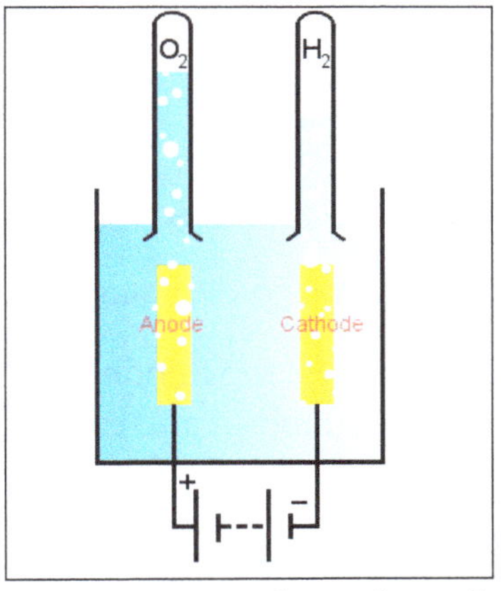

Direct current can be used to split water, as shown in the diagram

At the cathode, hydrogen ions gain electrons:

$$2H^+(aq) + 2e^- \rightarrow H_2(g)$$

At the anode, water loses electrons, forming oxygen and hydrogen ions:

$$2H_2O(l) \rightarrow O_2(g) + 4H^+(aq) + 4e^-$$

Half-Cell

A half-cell is half of an electrolytic or voltaic cell, where either oxidation or reduction occurs. The half-cell reaction at the anode is oxidation, while the half-cell reaction at the cathode is reduction.

Half Cell Reactions

For example, in the Zn–Cu battery, the two half cells make an oxidizing-reducing couple. Placing a piece of reactant in an electrolyte solution makes a half cell. Unless it is connected to another half cell via an electric conductor and salt bridge, no reaction will take place in a half cell.

On the cathode, reduction takes place.

$$Oxidant + ne^- \rightarrow Reductant$$

Example: $Cu_2^+ + 2e^- \rightarrow Cu$

Cu_2 + is the oxidizing agent and Cu the reducing agent.

On the anode, oxidation takes place.

$$Reductant \rightarrow Oxidant + ne^-$$

Example: $Zn \rightarrow Zn^{2+} + 2e^-$

Zn is the reducing agent, and Zn^{2+} the oxidizing agent.

A battery requires at least two electrodes, the anode at which oxidation occurs, and the cathode at which reduction occurs. Reduction and oxidation are always required in any battery setup.

A battery operation requires an anode, a cathode, a load, and a salt bridge (if the salt bridge is not there already). These are the key elements of a battery.

Nernst Equation

Nernst equation is a general equation that relates the Gibbs free energy and cell potential in electrochemistry. It is very helpful in determining cell potential, equilibrium constant etc.

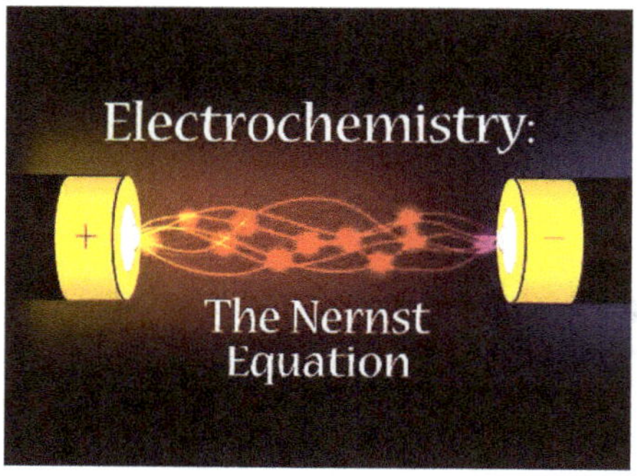

It takes into account the values of standard electrode potentials, temperature, activity and the reaction quotient for the calculation of cell potential. For any cell reaction, Gibbs free energy can be related to standard electrode potential as:

$$\Delta G = -nFE$$

Where, ΔG= Gibbs free energy, n = number of electrons transferred in the reaction, F = Faradays constant (96,500 C/mol) and E= cell potential. Under standard conditions, the above equation can be given as,

$$\Delta G^o = -nFE^o$$

According to the theory of thermodynamics, Gibbs free energy under general conditions can be related to Gibbs free energy under the standard condition and the reaction quotient as:

$$\Delta G = \Delta G^o + RT\ lnQ$$

Where, Q= reaction quotient, R= universal gas constant and T= temperature in Kelvin. Incorporating the value of ΔG and ΔG^o from the first two equations, we get:

$$-nFE = -nFE^0 + RT\ lnQ$$
$$E = E^0 - (RT/nF)\ lnQ$$

Converting natural log to log10, the above equation is known as the Nernst equation. Here, it relates the reaction quotient and the cell potential. Special cases of Nernst equation:

$$E = E^o - (2.303RT/nF) log_{10}Q$$

At standard temperature $T = 298\ K$:

$$E = E^o - (0.0592V/n)\ log_{10}Q$$

At standard temperature $T = 298\ K$, the $2.303RTF$, term equals $0.0592\ V$.

Under Equilibrium Condition

As the redox reaction in the cell proceeds, the concentration of reactants decreases while the concentration of products increases. This goes on until equilibrium is achieved. At equilibrium, $\Delta G = 0$. Hence, cell potential, $E = 0$.

Thus, the Nernst equation can be modified to:

$$E^0 - (2.303RT/nF)\ log_{10}Keq = 0$$
$$E^0 = (2.303RT/nF)\ log_{10}Keq$$

Where, Keq = equilibrium constant and F= faradays constant. Thus, the above equation gives us a relation between standard electrode potential of the cell in which the reaction is taking place and the equilibrium constant.

Redox Reaction

An oxidation-reduction (redox) reaction is a type of chemical reaction that involves a transfer of electrons between two species. An oxidation-reduction reaction is any chemical reaction in which the oxidation number of a molecule, atom, or ion changes by gaining or losing an electron. Redox reactions are common and vital to some of the basic functions of life, including photosynthesis, respiration, combustion, and corrosion or rusting.

References

- Basics_of_Electrochemistry, Electrochemistry, Analytical_Chemistry: libretexts.org, Retrieved 12 March, 2019
- Electrochemical-cells, boundless-chemistry: lumenlearning.com, Retrieved 16 June, 2019
- Electrochemical-cell: chemicool.com, Retrieved 26 April, 2019
- Galvanic-cell-definition: thoughtco.com, Retrieved 29 June, 2019
- Electrolytic-cell: chemicool.com, Retrieved 9 January, 2019
- Definition-of-half-cell: thoughtco.com, Retrieved 13 July, 2019
- Half-Cell_Reaction, Electrochemistry, Basics_of_Electrochemistry, Analytical_Chemistry: libretexts.org, Retrieved 11 August, 2019
- Nernst-equation, electrochemistry, chemistry: toppr.com, Retrieved 10 February, 2019
- Oxidation-Reduction_Reactions, Redox_Chemistry, Electrochemistry, Analytical_Chemistry: libretexts.org, Retrieved 18 May, 2019

8

Acids and Bases

An acid is a molecule or an ion that can donate a proton as well as form a covalent bond with an electron pair. Bases are the substance that release hydroxide in an aqueous solution. ph is the scale that is used to identify acidic or basic nature of a water solution. This chapter has been carefully written to provide an easy understanding of the varied facets of acids and bases.

There are several methods of defining acids and bases. While these definitions don't contradict each other, they do vary in how inclusive they are. The most common definitions of acids and bases are Arrhenius acids and bases, Brønsted-Lowry acids and bases, and Lewis acids and bases. Antoine Lavoisier, Humphry Davy, and Justus Liebig also made observations regarding acids and bases, but didn't formalize definitions.

Svante Arrhenius Acids and Bases

The Arrhenius theory of acids and bases dates back to 1884, building on his observation that salts, such as sodium chloride, dissociate into what he termed ions when placed into water.

- Acids produce H^+ ions in aqueous solutions;
- Bases produce OH^- ions in aqueous solutions;
- Water required, so only allows for aqueous solutions;
- Only protic acids are allowed; required to produce hydrogen ions;
- Only hydroxide bases are allowed.

Johannes Nicolaus Brønsted - Thomas Martin Lowry Acids and Bases

The Brønsted or Brønsted-Lowry theory describes acid-base reactions as an acid releasing a proton and a base accepting a proton. While the acid definition is pretty much the same as that proposed by Arrhenius (a hydrogen ion is a proton), the definition of what constitutes a base is much broader.

- Acids are proton donors;
- Bases are proton acceptors;

- Aqueous solutions are permissible;
- Bases besides hydroxides are permissible;
- Only protic acids are allowed.

Gilbert Newton Lewis Acids and Bases

The Lewis theory of acids and bases is the least restrictive model. It doesn't deal with protons at all, but deals exclusively with electron pairs.

- Acids are electron pair acceptors;
- Bases are electron pair donors;
- Least restrictive of the acid-base definitions.

Properties of Acids and Bases

Robert Boyle described the qualities of acids and bases in 1661. These characteristics may be used to easily distinguish between the two sets up chemicals without performing complicated tests:

Acids

- Taste sour (don't taste them)—the word 'acid' comes from the Latin acere, which means 'sour';
- Acids are corrosive;
- Acids change litmus (a blue vegetable dye) from blue to red;
- Their aqueous (water) solutions conduct electric current (are electrolytes);
- React with bases to form salts and water;
- Evolve hydrogen gas (H_2) upon reaction with an active metal (such as alkali metals, alkaline earth metals, zinc, aluminum).

Common Acids

- Citric acid (from certain fruits and veggies, notably citrus fruits);
- Ascorbic acid (vitamin C, as from certain fruits);
- Vinegar (5% acetic acid);
- Carbonic acid (for carbonation of soft drinks);
- Lactic acid (in buttermilk).

Bases

- Taste bitter (don't taste them);
- Feel slippery or soapy (don't arbitrarily touch them);
- Bases don't change the color of litmus; they can turn red (acidified) litmus back to blue;
- Their aqueous (water) solutions conduct an electric current (are electrolytes);
- React with acids to form salts and water.

Common Bases

- Detergents;
- Soap;
- Lye (NaOH);
- Household ammonia (aqueous).

Strong and Weak Acids and Bases

The strength of acids and bases depends on their ability to dissociate or break into their ions in water. A strong acid or strong base completely dissociates (e.g., HCl or NaOH), while a weak acid or weak base only partially dissociates (e.g., acetic acid).

The acid dissociation constant and base dissociation constant indicates the relative strength of an acid or base. The acid dissociation constant Ka is the equilibrium constant of an acid-base dissociation:

$$HA + H_2O \leftrightarrows A^- + H^3O^+$$

Where HA is the acid and A^- is the conjugate base.

$$K_a = \left[A^-\right]\left[H_3O^+\right] / [HA][H_2O]$$

This is used to calculate pK_a, the logarithmic constant:

$$pk_a = -\log 10\ K_a$$

The larger the pK_a value, the smaller the dissociation of the acid and the weaker the acid. Strong acids have a pK_a of less than -2.

pH

pH is a measure of hydrogen ion concentration, a measure of the acidity or alkalinity of a solution. The pH scale usually ranges from 0 to 14. Aqueous solutions at 25 °C with a pH less than 7 are acidic, while those with a pH greater than 7 are basic or alkaline. A pH level of 7.0 at 25 °C is defined as "neutral" because the concentration of H_3O^+ equals the concentration of OH^- in pure water. Very strong acids might have a negative pH, while very strong bases might have a pH greater than 14.

pH Equation

The equation for calculating pH was proposed in 1909 by Danish biochemist Søren Peter Lauritz Sørensen:

$$pH = -\log\left[H^+\right]$$

Where log is the base-10 logarithm and $[H^+]$ stands for the hydrogen ion concentration in units of moles per liter solution. The term "pH" comes from the German word "potenz," which means "power," combined with H, the element symbol for hydrogen, so pH is an abbreviation for "power of hydrogen."

Examples of pH Values of Common Chemicals

We work with many acids (low pH) and bases (high pH) every day. Examples of pH values of lab chemicals and household products include:

- 0: hydrochloric acid;
- 2.0: lemon juice;
- 2.2: vinegar;
- 4.0: wine;
- 7.0: pure water (neutral);
- 7.4: human blood;
- 13.0: lye;
- 14.0: sodium hydroxide.

Not All Liquids Have a pH Value:

pH only has meaning in an aqueous solution (in water). Many chemicals, including liquids, do not have pH values. If there's no water, there's no pH. For example, there is no pH value for vegetable oil, gasoline, or pure alcohol.

pH Scale

The pH scale is alternatively sometimes called the pH-acid-base scale and sometimes just the acid-base scale.

In neutral solutions, i.e. those that are neither acidic nor basic, $pH = 7.0$.

Acidic solutions are those with pH less than 7, while basic solutions have pH greater than 7.

A solution is:

- Acidic when there are more H^+ ions than OH^- ions present;
- Basic when there are more OH^- ions than H^+ ions present;
- Neutral when there are equal number of H^+ and OH^- ions present.

pH is a measure of the hydrogen ion activity, typically in aqueous solution.

In pH, the p stands for -log10 and the H stands for hydrogen ion activity:

$$pH = -log_{10}a_{H^+}$$

In dilute solutions, the activity of hydrogen ions can be approximated by their molar concentration, and it is in this form that pH is normally used:

$$pH = -log_{10}\left[H^+\right]$$

Why is pH=7 Neutral in Aqueous Solutions?

To see why this should be so, consider the following:

Water exists as an equilibrium between molecules and ions,

$$H_2O \rightleftharpoons H^+ + OH^-$$

The multiplication product of the concentrations $[H^+]$ and $[OH^-]$ is a constant; it is water's dissociation constant, which has a known value of 10-14 M_2.

$$K_w = \left[H^+\right]\left[OH^-\right] = 10^{-14}\ M^2$$

For neutrality, $[H^+]$ must be equal to $[OH^-]$. This means these quantities must both equal 10-7 M.

If we put $\left[H^+\right] = 10-7\ M$ into the equation for pH:

$$pH = -log_{10}\left[H^+\right]$$

We get pH = 7 for neutrality.

Acid and Base pH

Acid solutions $pH < 7$. pH values lower than 0 are possible.

Basic solutions have $pH < 7$. pH values greater than 14 are possible.

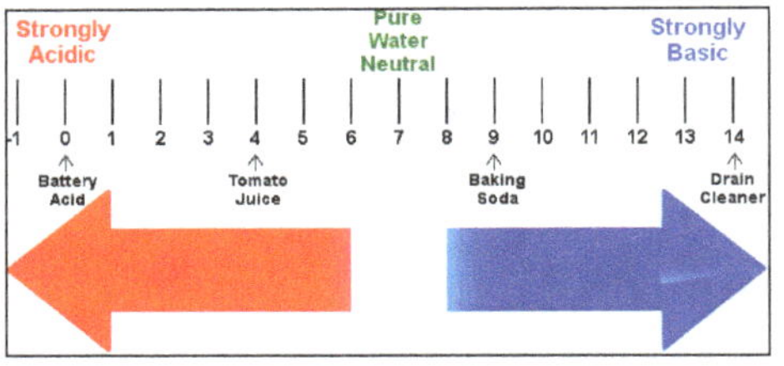

The pH Scale

pH and pKa

Once you have pH or pKa values, you know certain things about a solution and how it compares with other solutions:

- The lower the pH, the higher the concentration of hydrogen ions, $[H^+]$. The lower the pK_a, the stronger the acid and the greater its ability to donate protons.
- pH depends on the concentration of the solution. This is important because it means a weak acid could actually have a lower pH than a diluted strong acid. For example, concentrated vinegar (acetic acid, which is a weak acid) could have a lower pH than a dilute solution of hydrochloric acid (a strong acid). On the other hand, the pK_a value is a constant for each type of molecule. It is unaffected by concentration.
- Even a chemical ordinarily considered a base can have a pK_a value because the terms "acids" and "bases" simply refer to whether a species will give up protons (acid) or remove them (base). For example, if you have a base Y with a pK_a of 13, it will accept protons and form YH, but when the pH exceeds 13, YH will be deprotonated and become Y. Because Y removes protons at a pH greater than the pH of neutral water, it is considered a base.

Henderson-Hasselbalch Equation

The Henderson-Hasselbalch equation relates pH, pK_a, and molar concentration (concentration in units of moles per liter):

$$pH = pK_a + \log\left([A^-]/[HA]\right)$$

$[A^-]$ = molar concentration of a conjugate base

$[HA]$ = molar concentration of an undissociated weak acid (M)

The equation can be rewritten to solve for pOH:

$$pOH = pK_b + log\left([HB^+]/[B]\right)$$

$[HB^+]$ = molar concentration of the conjugate base (M)

$[B]$ = molar concentration of a weak base (M)

Buffers and Indicators

Buffers are a type of solution that resist a change in pH when an acid or base is added to it. Buffer solutions usually contain a weak acid or base combined with its conjugate acid or base. As a result, they can also be referred to as solutions that are simply responsible for combining an acid and a base. In addition, they are able to neutralize fairly small amounts of added acid and base, which results in it being able to maintain a stable pH in a solution. Buffers are especially important for reactions or processes that need a specific and stable pH. They have a working pH range and capacity that can dictate how much acid and base can be neutralized before the pH changes, and the specific amount by which it will change. Below are some examples dictating what happens in a solution with the presence of a buffer:

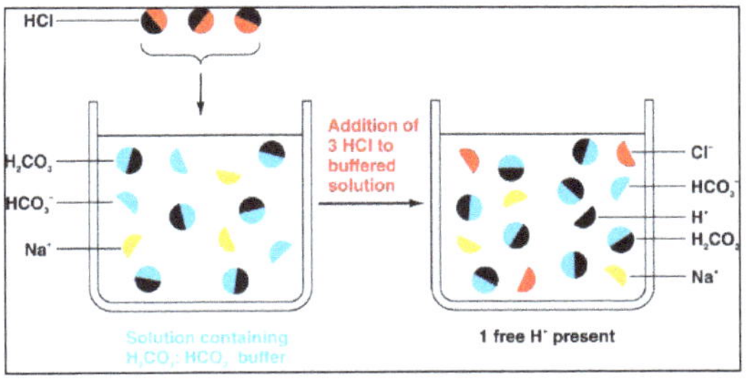

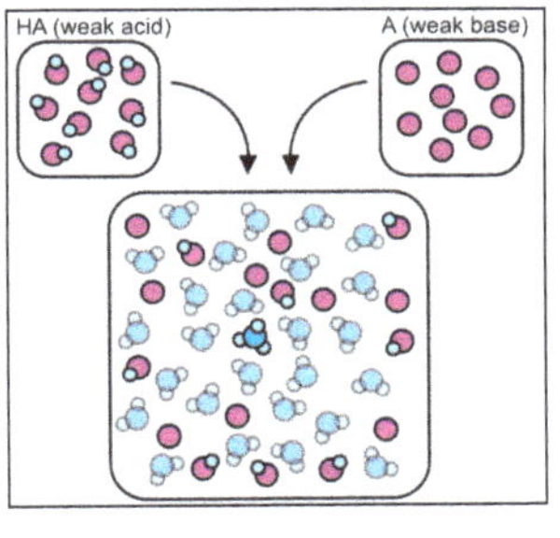

Buffers can also be found in many of the things that we use in everyday life.

- The blood in our bodies has a constant pH of 7.4 thanks to buffers in our bodies;
- Buffers are also used in our bodies in the hemoglobin complex, which helps the body utilize oxygen better;
- The making of many alcoholic beverages requires buffers (related to fermentation);
- Buffers are used in the production of shampoos (examples include citric acid and sodium hydroxide);
- Buffers are used in baby powder, which helps inhibit the growth of bacteria by maintaining a pH of 6.

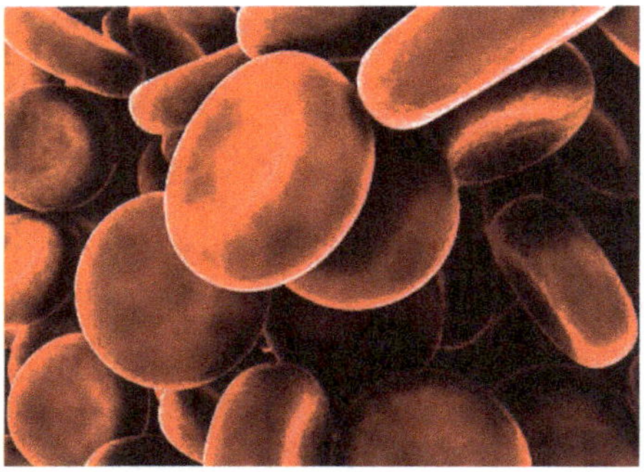

An indicator is a compound that has the ability to change color at a specific pH value or in the presence of a particular substance. Indicators can be used to monitor the acidity or alkalinity of a solution and the progress of a reaction. The indicator itself is typically a weak acid or base. An indicator does not change color from pure acid to pure base/ alkaline at a specific hydrogen ion concentration, but instead changes over a range of hydrogen ion concentrations. This specific range is called the color change interval, and is expressed as a pH range. There are many different types of indicators.

Hydrolysis

Hydrolysis is a type of decomposition reaction where one reactant is water. Typically, water is used to break chemical bonds in the other reactant. The term comes from the Greek prefix hydro- (meaning water) with lysis (meaning to break apart).

Hydrolysis may be considered the reverse of a condensation reaction, in which two molecules combine with each other, producing water as one of the products.

The general formula of a hydrolysis reaction is:

$$AB + H_2O \rightarrow AH + BOH$$

Organic hydrolysis reactions involve the reaction of water and an ester. This reaction follows the general formula:

$$RCO-OR' + H_2O \rightarrow RCO-OH + R'-OH$$

The hyphen on the left side denotes the covalent bond that is broken during the reaction.

Hydrolysis Examples

Dissolving a salt of a weak acid or base in water is an example of a hydrolysis reaction. Strong acids may also be hydrolyzed. For example, dissolving sulfuric acid in water yields hydronium and bisulfate.

Hydrolysis of a sugar has its own name: saccharification. For example, the sugar sucrose may undergo hydrolysis to break into its component sugars, glucose and fructose.

Acid-base catalyzed hydrolysis is another type of hydrolysis reaction. An example is the hydrolysis of amides.

Acid-base Reactions

The reaction between an acid and a base is known as a neutralisation reaction. Often when an acid and base react a salt and water will be formed. We will look at a few examples of acid-base reactions.

- Hydrochloric acid reacts with sodium hydroxide to form sodium chloride (a salt) and water. Sodium chloride is made up of Na^+ cations from the base (NaOH) and Cl^- anions from the acid (HCl).

  $$HCl\ (aq) + NaOH\ (aq) \rightarrow H_2O\ (l) + NaCl\ (aq)$$

- Hydrogen bromide reacts with potassium hydroxide to form potassium bromide (a salt) and water. Potassium bromide is made up of K^+ cations from the base (KOH) and Br^- anions from the acid (HBr).

  $$HBr\ (aq) + KOH\ (aq) \rightarrow H_2O\ (l) + KBr\ (aq)$$

- Hydrochloric acid reacts with ammonia to form ammonium chloride (a salt). Ammonium chloride is made up of NH_4^+ cations from the base (NH_3) and Cl^- anions from the acid (HCl).

  $$HCl\ (aq) + NH_3(aq) \rightarrow NH_4Cl\ (aq)$$

References

- Acids-and-bases-definitions: thoughtco.com, Retrieved 19 April, 2019
- Definition-of-ph-in-chemistry: thoughtco.com, Retrieved 21 March, 2019
- The-ph-and-pka-relationship: thoughtco.com, Retrieved 23 January, 2019
- Henderson-hasselbalch-equation-and-example: thoughtco.com, Retrieved 3 June, 2019
- Definition-of-hydrolysis: thoughtco.com, Retrieved 6 May, 2019
- Types-of-reactions: siyavula.com, Retrieved 17 February, 2019

Permissions

All chapters in this book are published with permission under the Creative Commons Attribution Share Alike License or equivalent. Every chapter published in this book has been scrutinized by our experts. Their significance has been extensively debated. The topics covered herein carry significant information for a comprehensive understanding. They may even be implemented as practical applications or may be referred to as a beginning point for further studies.

We would like to thank the editorial team for lending their expertise to make the book truly unique. They have played a crucial role in the development of this book. Without their invaluable contributions this book wouldn't have been possible. They have made vital efforts to compile up to date information on the varied aspects of this subject to make this book a valuable addition to the collection of many professionals and students.

This book was conceptualized with the vision of imparting up-to-date and integrated information in this field. To ensure the same, a matchless editorial board was set up. Every individual on the board went through rigorous rounds of assessment to prove their worth. After which they invested a large part of their time researching and compiling the most relevant data for our readers.

The editorial board has been involved in producing this book since its inception. They have spent rigorous hours researching and exploring the diverse topics which have resulted in the successful publishing of this book. They have passed on their knowledge of decades through this book. To expedite this challenging task, the publisher supported the team at every step. A small team of assistant editors was also appointed to further simplify the editing procedure and attain best results for the readers.

Apart from the editorial board, the designing team has also invested a significant amount of their time in understanding the subject and creating the most relevant covers. They scrutinized every image to scout for the most suitable representation of the subject and create an appropriate cover for the book.

The publishing team has been an ardent support to the editorial, designing and production team. Their endless efforts to recruit the best for this project, has resulted in the accomplishment of this book. They are a veteran in the field of academics and their pool of knowledge is as vast as their experience in printing. Their expertise and guidance has proved useful at every step. Their uncompromising quality standards have made this book an exceptional effort. Their encouragement from time to time has been an inspiration for everyone.

The publisher and the editorial board hope that this book will prove to be a valuable piece of knowledge for students, practitioners and scholars across the globe.

Index

P

R

S

T

V

Z

www.ingramcontent.com/pod-product-compliance
Lightning Source LLC
LaVergne TN
LVHW080602210926
848485LV00003B/94
* 9 7 8 1 6 3 2 3 8 8 3 8 4 *